ADVANCES IN
GROUP PROCESSES

Volume 13 • 1996

ADVANCES IN GROUP PROCESSES

Volume Editors: BARRY MARKOVSKY
MICHAEL J. LOVAGLIA
ROBIN SIMON
Department of Sociology
University of Iowa

Series Editor: EDWARD J. LAWLER
Department of Organizational Behavior
Cornell University

VOLUME 13 • 1996

 JAI PRESS INC.

Greenwich, Connecticut *London, England*

CONTENTS

LIST OF CONTRIBUTORS

Nancy L. Eiesland

Candler School of Theology
Emory University

Rebecca Ford

Department of Sociology
University of Florida

Yuen J. Huo

Department of Psychology
University of California at Berkeley

Cathryn Johnson

Department of Sociology
Emory University

Roderick M. Kramer

Graduate School of Business
Stanford University

Michael J. Lovaglia

Department of Sociology
University of Iowa

Barry Markovsky

Department of Sociology
University of Iowa

Douglas Mason-Schrock

Department of Sociology
North Carolina State University

Linda D. Molm

Department of Sociology
University of Arizona

Jeylan T. Mortimer

Life Course Center
University of Minnesota

Brian Powell

Department of Sociology
Indiana University

Dawn T. Robinson

Department of Sociology
Louisiana State University

Michael L. Schwalbe

Department of Sociology
North Carolina State University

Michael J. Shanahan   Department of Human Development
                                   and Family Studies
                                   Pennsylvania State University

Robin Simon   Department of Sociology
                                   University of Iowa

Heather J. Smith   Department of Psychology
                                   University of California at Berkeley

Lala Carr Steelman   Department of Sociology
                                   University of South Carolina

Tom R. Tyler   Department of Psychology
                                   University of California at Berkeley

PREFACE

EDITORIAL POLICY

Advances in Group Processes publishes theoretical analyses, reviews, and theory-based empirical chapters on group phenomena. The series adopts a broad conception of "group processes." This includes work on groups ranging from the very small to the very large and on classic and contemporary topics such as status, power, exchange, justice, influence, decision making, intergroup relations, and social networks. Contributors have included scholars from diverse fields including sociology, psychology, political science, and organizational behavior.

The series provides an outlet for papers that may be longer, more theoretical, and/or more integrative than those published by standard journals. We place a premium on the development of testable theories and on theory-driven research programs. Chapters in the following categories are especially apropos:

- *Conventional and unconventional theoretical work, from broad metatheoretical and conceptual analyses to refinements of existing theories and hypotheses.* One goal of the series is to advance the field of group processes by promoting theoretical work.
- *Papers that review and integrate programs of research.* The current structure of the field often leads to the piecemeal publication of different parts of a program of research. This series offers those engaged in programmatic

research on a given topic an opportunity to integrate their published and unpublished work into a single paper. Review articles that transcend the author's own work are also of considerable interest.

- *Papers that develop and apply social psychological theories and research to macrosociological processes.* One premise underlying this series is that links between macro- and microsociological processes warrant more systematic and testable theorizing. The series encourages development of the macrosociological implications embedded in social psychological work on groups.

In addition, the editors are open to submissions that depart from these guidelines.

CONTENTS OF VOLUME 13

Exemplifying the vitality and breadth of work in the field of group processes, the nine chapters in this volume offer new theories, new research, new integrations, and new perspectives. Below we provide brief summaries of each contribution. These thumbnail sketches cannot do justice to the chapters that follow, so we hope that your interest will be piqued and that you will read on. You will be well rewarded for doing so.

In group processes research, it has been said that a good way to make a name for oneself is to demonstrate that a phenomenon which "everyone knows" results from a personality trait actually depends on some aspect of social structure. Kramer's chapter ("Paranoia, Distrust, and Suspicion Within Social Groups: A Social Categorization Perspective") does just that. His chapter develops a theory of how paranoia is socially engendered. His studies of M.B.A. students show how their token status and tenure in the group affect their level of distrust and suspicion, and how paranoid cognitions can lead to a variety of socially maladaptive behaviors.

The chapter by Tyler, Smith, and Huo ("Member Diversity and Leadership Effectiveness: Procedural Justice, Social Identity, and Group Dynamics") uses group processes to explain how diverse members of a large society can function effectively together. Their theory proposes that, although actors are in large part self-interested, procedural justice can bridge differences between subgroups. Thus, for instance, by instituting just procedures, diverse groups may perceive authorities as legitimate even though inequality in wealth among groups remains. The theory is supported by studies using a variety of empirical techniques.

Eiesland and Johnson ("Physical Ability as a Diffuse Status Characteristic: Implications for Small Group Interaction") assert that physical ability operates as a "diffuse status characteristic" in American society. From the standpoint of the theory of status characteristics and expectation states, this means that physical ability has a variety of implications for mutual expectations of task performance and for interaction patterns in task groups. The authors offer preliminary evidence for these processes and provide some strategies for reducing the undesirable aspects of status generalization based on physical ability.

The chapter by Robinson ("Identity and Friendship: Affective Dynamics and Network Formation") applies affect control theory to the problem of how friends are chosen. The theory holds that individuals seek to maintain identities and to control certain fundamental sentiments. Friendship choices may then be viewed as opportunities to enact those aspects of one's identity that are positively valued. Robinson uses a substantively rich computer simulation technique to show relationships between the affect control process and emergent patterns of friendship formation in groups.

Another chapter that builds on the social identity theme is Schwalbe and Mason-Schrock's "Identity Work as Group Process." They extend previous theory on identity creation by proposing that identity work is not done by individuals alone but in concert with others as part of a group process. They provide a conceptual sketch of the process of subcultural identity work, draw on ethnographic material to illustrate their theoretical constructs, and show how subcultural identity work can be understood in cultural and structural contexts.

Molm's chapter ("Punishment and Coercion in Social Exchange") integrates a vast amount of research to develop a theory that distinguishes punishment or coercive power from reward power in social exchange. Actors are loss averse, meaning that while punishment is potentially effective, it involves much greater subjective risk than does reward. She also points out that Emerson's dictum, "To have power is to use it," does not necessarily apply to punishment. The most recent theoretical advances predict differences in the relative strength of reward vs. punishment power at varying levels of interdependence.

Shanahan and Mortimer ("Understanding the Positive Consequences of Psychosocial Stressors") ask what social psychological mechanisms facilitate adaptation to stressors and thereby reduce distressful reactions. Whereas the lion's share of stress research concerns deleterious outcomes, in this chapter the authors elucidate a process whereby a variety of positive outcomes, known collectively as "eustress," follow from the onset of social stressors. The approach is sufficiently general so as to accommodate differences in individual responses to the same stressful situations. These differences are conditional on such factors as life histories, self-esteem, and self-efficacy.

Steelman and Powell's provocative chapter ("The Family Devalued: The Treatment of the Family in Small Groups Literature") argues that despite numerous points of overlap between family research and small groups research, there has been little cross-fertilization between these two lines of inquiry. They review and synthesize research on the link between family structure and educational outcomes of children, discuss various reasons for the intellectual division between family and small groups research, and advance some ways to bridge the gap between these two lines of inquiry.

Only infrequently do participants in scientific research programs step back and examine some of the broader implications of their choice of research methods. Rebecca Ford's contribution addresses this issue and others by examining the mutual impact of theory and method in a program of experimental research. "Metamethodology and Procedural Decisions in Theoretical Experimentation: The Case of Power Processes in Bargaining" traces the web of connections between the elements of metatheory, theory, and scope conditions, and critical decisions regarding what variables are manipulated and what variables are controlled in tests of the theory.

We hope that you enjoy these contributions and that you find them to be as interesting and valuable as we have.

Barry Markovsky
Michael J. Lovaglia
Robin Simon
Volume Co-Editors

PARANOIA, DISTRUST, AND SUSPICION WITHIN SOCIAL GROUPS:
A SOCIAL CATEGORIZATION PERSPECTIVE

Roderick M. Kramer

ABSTRACT

This paper presents a framework for conceptualizing antecedents and consequences of distrust and suspicion within diverse social groups. Drawing on recent theory and research on paranoid social cognition and social categorization, this framework explicates how individuals' perceptions of being distinctive or different from other group members affects social information processing, including judgments about trust and distrust. Two laboratory experiments are presented that investigate hypotheses derived from this framework. These experiments examine how two important dimensions along which group members are sometimes distinctive from others (token status and newcomer status) affect processing of trust-related information. The results document how tokens' and newcomers' perceptions of distrust and suspicion differ from the perceptions of nontokens and those with longer tenure in the group.

Advances in Group Processes, Volume 13, pages 1-32.
Copyright © 1996 by JAI Press Inc.
All rights of reproduction in any form reserved.
ISBN: 0-7623-0005-1

I get the willies whenever I see closed doors. Even at work, where I am doing so well now, the sight of a closed door is sometimes enough to make me dread that something horrible is happening behind it, something that is going to affect me adversely.

—Joseph Heller, *Something Happened* (1966, p. 1)

The study of social perception in groups is currently enjoying a considerable renaissance. This reawakening is due, at least in part, to steady progress over the past decade in the development of sophisticated social cognitive theories and methods which can be brought to bear on the exploration of group phenomena (e.g., Mullens and Goethals 1987). One seemingly robust conclusion emerging from this research is that individuals often maintain a variety of relatively positive beliefs about the social groups to which they belong. For example, research has shown that individuals frequently enter groups with positive expectations about the benefits they will derive from group participation; they tend to attribute relatively positive traits to other ingroup members, perceiving them as honest, cooperative, and trustworthy—especially when compared to their perception of "outgroup" members; and they often give other group members the benefit of the doubt when trying to explain unattractive behaviors, while often making far less charitable attributions when explaining comparable behaviors by outgroup members (see, e.g., Brewer 1979; Brinthaupt, Moreland, and Levine 1991; Crocker and Luhtanen 1990; Janis 1983; Paulus, Dzindolet, Poletes, and Camacho 1993; Polzer, Kramer, and Neale 1995).

As with any caricature of a complex stream of research, this benign portrayal of social perception and group life tells only part of the story. In many social groups, individuals do not hold such uniformly favorable views about the dispositions of other group members, nor they do entertain such generous presumptions about their motives when trying to explain their behavior. Instead, they often question whether they are being treated fairly and with dignity in their exchanges; they fear others are not cooperating fully with them; and they may harbor deep suspicions about the intentions of other group members (Bies 1987; Kanter 1977a; Kramer 1994; Lind and Tyler 1988; Martin 1993). Within such groups, problems of distrust and suspicion are likely to be pervasive, permeating the social relations among group members and undermining their attempts to cooperate with each other.

Despite its importance, the problem of distrust and suspicion within social groups has received surprisingly little systematic attention from researchers (cf. Golembiewski and McConkie 1975). The lack of research on this topic is especially striking when compared to the robust literatures on interpersonal (Rotter 1980) and intergroup (Brewer 1981) trust and distrust. Accordingly, a primary aim of the present chapter is to explore some of the antecedents and consequences of distrust and suspicion within social groups. Specifically, the chapter articulates a framework for conceptualizing a form of "irrational" distrust or suspicion within groups that resembles social paranoia. This group-based form of paranoia can be characterized as a tendency for group members to overattribute hostile intentions and malevolent motives to other group members, even in the absence of concrete evidence on which to base such attributions. In developing the notion of paranoid social cognition within groups, the present chapter integrates and blends theory and research findings from two distinct streams of research: social psychological research on paranoid cognition (Fenigstein and Vanable 1992; Kramer 1994, 1995) and organizational research on social categorization processes (Ashforth and Mael 1989; Baron and Pfeffer 1994; Kanter 1977a; Kramer 1991; Tsui, Egan, and O'Reilly 1992; Wharton 1992).

The basic conjecture advanced in this chapter is that social categorization processes often create and reinforce tendencies for certain group members to misconstrue and overconstrue social interactions in personalistic (i.e., self-referential or self-relevant) terms. Thus, even relatively innocent organizational encounters (e.g., the failure of another organizational member to return a greeting in the hallway) are viewed as signaling something of significance about the person's standing in the relationship or value in the organization. As a result of such overly personalistic construals, they draw inferences that lead them to feel they are being unfairly subjected to hostile scrutiny and are the recipients of malevolent or inequitable treatment.

To advance this general argument, the chapter is organized as follows. First, I define irrational distrust and suspicion, distinguishing these notions from other conceptions of distrust and suspicion found in the social science literature. I then suggest how social categorization processes within groups might contribute to such distrust and suspicion. Next, I offer some empirical findings that support these arguments. I conclude by discussing a number of implications of the framework for theory and research on groups.

CONCEPTUALIZING DISTRUST AND SUSPICION WITHIN GROUPS: THE ROLE OF PARANOID SOCIAL COGNITION

To develop the concept of group paranoia and show how it contributes to distrust and suspicion among group members, it may be helpful first to offer a few preliminary remarks regarding how distrust and suspicion have been conceptualized in previous theory and research.

Overview of Previous Perspectives

Previous theory and research has identified a large number of psychological and social processes that contribute to distrust and suspicion (see, e.g., Brewer 1981;

Deutsch 1973; Fox 1974; Gambetta 1987; Granovetter 1985; Luhman 1979; Sitkin and Roth 1993; Wrightsman 1991; Zucker 1986). Psychological research on individual differences in the propensity to distrust others (e.g., Gurtman 1992; Rotter 1967; Wrightsman 1981) has shown that distrust and suspicion are correlated with a variety of general attitudes and beliefs about other people. For example, Gurtman (1992) demonstrated that individuals who are high in Machiavellian orientation are more likely to be distrusting and suspicious of other people, compared to individuals who are low in this orientation. Behaviorally-oriented research has shown further that distrust and suspicion are influenced by the history of interaction between individuals, resulting in generalized expectancies regarding their trustworthiness or lack of trustworthiness (Deutsch 1973; Lindskold 1978; Rotter 1980). Other studies have documented how social ties and structural relations linking interdependent actors influence the level of distrust between them (Brewer 1981; Granovetter 1985; Zucker 1986).

Although these literatures differ considerably in the extent to which they emphasize micro- vs. macro-level determinants of distrust and suspicion, several points of convergence are nonetheless discernible across them. First is the assumption that distrust and suspicion can be conceptualized as psychological states that are closely linked to the attributions individuals make about other people's behavior. In particular, distrust and suspicion arise when individuals attribute hostile motives or untrustworthy intentions to others' actions, especially in situations where uncertainty or ambiguity is present regarding the cause of their behavior (see, e.g., Deutsch 1973; Lindskold 1978).

These attributional processes have often been construed as reasonably rational and orderly forms of social inference, consistent with the idea that social perceivers resemble "intuitive scientists" trying to make sense of the social and organizational worlds they inhabit (Kelley 1973). For example, Rotter (1980) and Lindskold (1978) conceptualized distrust as a generalized expectancy or belief regarding the lack of trustworthiness of other individuals that is predicated upon the history of interaction with them. According to this view, when people make judgments about others' trustworthiness (or lack of it), they act much like Bayesian statisticians whose inferences are continually updated on the basis of recent experiences. Research in this tradition has shed considerable light on the conditions under which this kind of distrust evolves. For example, it has been closely linked to patterns of interaction or exchange that involve repeated violations of reciprocity (Deutsch 1973; Lindskold 1978; Rotter 1980). According to these models, when individuals act on faith only to discover subsequently that faith has been abused, their trust in the other declines. Distrust and suspicion of this sort are rational in that they are tied to a concrete history that legitimates or justifies the revision of social expectations.

While recognizing the importance of these relatively rational forms of distrust, a number of researchers have noted that other forms of distrust and suspicion appear to be far less rational in their antecedents and origins (e.g., Barber 1983;

Deutsch 1973; Luhman 1979). For example, research on ethnocentrism and ingroup bias suggests that individuals sometimes distrust members of outgroups even in the absence of any history of previous interaction or contact (Brewer 1981). Similarly, research on personality correlates of distrust and suspicion suggests that some individuals seem predisposed to adopt an orientation of distrust toward others (Gurtman 1992). Deutsch (1973) characterized such irrational forms of distrust as an "inflexible, rigid, unaltering tendency to act in a suspicious manner, *irrespective of the situation* or the consequences of so acting." He goes on to add, "The pathology [of this type of distrust] is reflected in the indiscriminateness and incorrigibility of the behavioral tendency" (p. 171). Defined in these terms, irrational distrust represents a form of presumptive distrust that is conferred ex ante on other social actors. In this sense, it reflects an exaggerated propensity towards distrust, which can arise even in the absence of specific experiences that justify or warrant it.

While the notion of irrational distrust is intriguing, we know little about its origins or consequences, especially in the context of social and organizational groups. In the following section, therefore, I attempt to address some of these issues.

Paranoid Cognition: Social Inference Gone Awry

As a starting point for analysis, it is useful to think about group members' judgments about distrust and suspicion from a purely social cognitive perspective. As characterized by Fiske and Taylor (1991), the study of social cognition concerns" how people make sense of other people and themselves" (p. 14). Social cognition theory and research attempts to (1) clarify the role cognitive processes play in social perception and judgment and (2) indicate how such perceptions and judgments, in turn, influence social behavior and interaction. When applied to group phenomenon, social cognitions can be construed as individuals' cognitions about (1) other members of a group to which they belong and /or (2) their cognitions about the group as a whole.

Paranoid cognitions represent a special class of social cognition characterized by the *mis*attribution and *over*attribution of malevolent motives and sinister intentions to others' behavior (Fenigstein and Vanable 1992; Kramer 1994). They have been defined formally as "persecutory delusions and false beliefs whose propositional content clusters around ideas of being harassed, threatened, harmed, subjugated, persecuted, accused, mistreated, wronged, tormented, disparaged, vilified ... by malevolent others, either specific individuals or groups" (Colby 1981, p. 518).

Within the field of clinical psychology, paranoid cognitions have been regarded usually as symptomatic of an individual psychiatric disorder (American Psychological Association 1987; Cameron 1943). Hence, the emphasis on delusion and false belief in the formal definition just offered. This disorder is presumed to be caused by abnormal personality dynamics that generate persistant and pervasive inferential errors. Recent social cognitive research, however, has advanced a very different conception of paranoid cognitions, and one that affords considerably more attention to their social and situational determinants (see, in particular, Fenigstein and Vanable 1992; Kramer 1994, 1995; Zimbardo, Andersen, and Kabat 1981). This latter conception proceeds from the intuition that, in milder form, paranoid cognitions appear to be quite prevalent and observed even among "normal" individuals, especially individuals who find themselves in certain kinds of awkward or threatening social situations. As Fenigstein and Vanable (1992) cogently observe in this regard, ordinary people "in their everday behavior often manifest characteristics—such as self-centered thought, suspiciousness, assumptions of ill will or hostility, and even notions of conspiratorial intent—that are reminiscent of paranoia ... on various occasions, one may think one is being talked about or feel as if everything is going against one, resulting in suspicion and mistrust of others, as though they were taking advantage of one or were to blame for one's difficulties" (p. 130-133).

In addition to suggesting that paranoid cognitions are fairly common, this research has shed considerable light on the antecedents of paranoid social cognition. Of particular relevance to the present analysis is evidence that paranoid cognitions are likely to occur in situations where individuals feel (1) a heightened sense of self-consciousness, (2) under intense evaluative scrutiny, and /or (3) that their personal identity or self-esteem is threatened.

This research raises several questions. First, under what circumstances might individuals within a social group feel a heightened sense of self-consciousness or under intense evaluative scrutiny by other group members? Second, when might group members feel their self-esteem or identity is threatened? In the next section, I offer one approach to these questions that is motivated by social categorization theory. According to this theory, individuals often define themselves and their relationships to other people in terms of the social categories which are salient during social interaction (see Ashforth and Mael 1989; Kramer 1991; Turner 1987 for recent overviews). Drawing on this theory, I argue that categorization processes in groups often generate the kind of heightened self-consciousness and perception of being under evaluative scrutiny that foster the emergence of paranoid cognitions.

SOCIAL CATEGORIZATION AND PARANOID SOCIAL COGNITION: CONCEPTUAL LINKS

All human beings possess membership in multiple social groups. As a consequence, they can categorize themselves—and be categorized by others in turn—in a variety of different ways. These include categorizations based upon physical attributes such as age, race, or gender; categorizations based upon socially-defined categories such as religion and social class; and organizationally-defined attributes, such as institutional affiliation, departmental membership, and job title. Recognizing their importance, researchers have afforded a great deal of attention in recent years to exploring how such categorization processes influence social perception within groups and organizations (e.g., Ashforth and Mael 1989; Baron and Pfeffer 1994; Kanter 1977a; Kramer 1991; Tsui, Egan, and O'Reilly 1992; Wharton 1992).

Several conclusions emerge from this research. First, social categorization can influence how individuals define themselves in a given social situation. Turner (1987) used the term "self-categorization" to refer to these processes and their effects. The self categorization process is affected by the particular social context within which individuals find themselves. In particular, there is evidence that individuals often tend to categorize themselves in terms of those attributes that are distinctive or unique in a given setting. For example, if an individual is the only female in a group, her distinctive gender status may be afforded disproportionate emphasis when explaining her behavior, affecting not only how she is seen by others but also how she sees herself as well. From a cognitive stand point, the distinctiveness of the category makes gender-based attributes more available during social information processing. As a result, they tend to become more salient or "loom larger" during social interaction, affecting both social inference processes and the behaviors that flow from them (Cota and Dion 1986; Kanter 1977a, 1977b; McGuire and Padawer-Singer 1976; Taylor 1981).

The impact of categorization processes on intragroup perception and behavior is highlighted by recent research showing that people attach considerable importance to information regarding their standing in the groups to which they belong (Lind and Tyler 1988). Standing refers to the "information communicated to a person about his or her status with the group ... *communicated both by interpersonal aspects of treatment—politeness and /or respect—and by the attention paid to a person as a full group member*" (Tyler 1993, p. 148, emphasis added). Because of their concern about standing, group members tend to use information regarding how they are treated during interactions and exchanges with other group members "diagnostically" (e.g., as clues to whether their stand ing is "good" or "bad"). In effect, individuals treat such information as if it is informative about such things as whether they will be treated fairly and whether they are valued by other members of the group or organization (Tyler 1993).

Because of their ongoing concerns about standing, individuals tend to be proactive information seekers, working hard to find "data" that will help them make sense of their place in the social order (Ashford and Cummings 1985; Morrison 1993). Unfortunately, in trying to do so, individuals are often torn between the desire to obtain *accurate* information about where they stand in a group and the desire to obtain reassuring or *self-enhancing* information (Greenwald 1980). As a result, when individuals find themselves in situations where their self-esteem or positive social identity is threatened or called into question, a conflict may arise between the desire to find out where they actually stand in the group and the goal of assuaging their fears and anxieties about poor standing or loss of standing.

I argue next that this conflict is often particularly acute for individuals who occupy certain social categories within groups. In particular, I suggest that individuals who belong to salient or distinctive social categories that are associated with ambiguity or uncertainty about standing are likely to suffer from a heightened self-consciousness. Because they feel that they are different, "stand out," or enjoy a distinctive status in the group, they tend to overestimate the extent to which they are under evaluative scrutiny by other group members. Such perceived scrutiny, and the self-consciousness it engenders in turn, foster a pattern of hypervigilant social information processing that leads such individuals to overattribute personalistic causes to others' actions, even when they are benign. As a consequence, categorization processes contribute to the development of specific group members' vulnerability to a paranoid style of social perception.

STUDY 1: THE LIABILITY OF BEING "DIFFERENT"

The argument that self-categorization on the basis of distinctive or exceptional status produces a heightened form of self-consciousness and perception of being under evaluative scrutiny is consistent with a considerable body of evidence regarding the cognitive and social consequences of merely "being different" from other members of a social group (Brewer 1991; Kanter 1977a, 1977b; Tsui, Egan, and O'Reilly 1992; Taylor 1981). Such self-consciousness might be expected to be especially pronounced when individuals are uncertain about precisely how much merely "being different" from others matters (i.e., its perceived diagnostic relevance with respect to questions or concerns about standing in a group). In social information processing terms, one fairly immediate consequence of individuals' awareness of "being different" from others is that it prompts effortful attributional search for the causes of others' behavior towards them, while at the same time injecting considerable ambiguity in the social inference process.

In many respects, so-called "token" members of social groups exemplify these cognitive dilemmas and difficulties. Token status in groups is based upon "ascribed characteristics (master statuses such as sex, race, religion, ethnic group, age, etc.) or other characteristics that carry with them a set of assumptions about culture, status, and behavior that are highly salient for majority category members" Kanter (1977b, p. 966).

In her now classic discussions of the effects of token status on social perception and interpersonal relations, Kanter (1977a, 1977b) notes that individuals who are members of token categories are likely to attract disproportionate attention from other groups members, particularly those who enjoy dominant status in terms of numerical proportions. For example, she observes that females in many American corporations often feel as if they are in the "limelight" compared to their statistically more numerous male counterparts.

In support of these observations, Taylor (1981) demonstrates that observers often allocate disproportionate amounts of attention to individuals who have token status in groups, especially when making attributions about group processes and outcomes. Lord and Saenz (1985) subsequently provide an important extension of this early work by showing how token status affects the cognitive processes of tokens themselves. Based on their findings, they conclude that, "Tokens feel the social pressure of imagined audience scrutiny, and may do so even when the 'audience' of majority group members treat them no differently from nontokens" (p. 919).

Extrapolating from these arguments, Study 1 investigated the hypothesis that group members who have token status will be more self-conscious and perceive themselves to be under evaluative scrutiny to a greater extent than nontoken group members. As a consequence, tokens will tend to overestimate the extent to which they are the targets of other's attention. This will cause them to over construe their interactions with other group members—especially interactions involving nontokens—in overly personalistic, self-referential terms.

Overview of Method

To investigate these hypotheses, I thought it would be helpful to go outside the laboratory. Laboratory studies of the effects of social categorization on group judgment and behavior have generally adopted the strategy of creating "minimal" social groups using arbitrary and transient manipulations of group identity. In contrast, I wanted to study social perception within an intact and enduring group—one in which membership was of real psychological significance to the individuals who belong to it. Accordingly, I chose to study incoming masters of business administration (M.B.A.) students at a major business school during the first term of their two-year program. The incoming M.B.A. classes in most major business schools are highly competitive groups. Individuals within such groups generally care a great deal about their relative standing or positional status with respect to academic performance. At the same time, they also care a great deal about their social standing in the group: students attach considerable value to fitting in and being accepted by the group. For example, M.B.A. students at Stanford and Harvard enter these groups believing that they are forming life long associations. Their classmates represent a potentially invaluable network of several hundred people on which they hope to draw throughout their professional life.

From the stand point of shaping individuals' perceptions of their token (or nontoken) status within this group, business schools are interesting organizations because, rather than conceal or de-emphasize category-based information, they often actively disseminate it. At the Stanford University Graduate School of Business, for example, detailed statistics as to the proportional representation of various social and organizational categories within the group are made available to each group member. The "diversity" of the group is a central topic during the socialization process and is frequently showcased in discussions about the rich organizational culture of the business school. Thus, individuals have precise information about how typical they are along a variety of categorical distinctions, including age, gender, educational attainment, the status of the undergraduate institutions they attended, and the previous employers for whom they worked.

Ostensibly, this information is provided to students in order to document the impressive level of diversity within the group, a characteristic that is highly valued by the organizational culture. Its psychological and social impact, however, may transcend such noble institutional motives. For example, publicity about diversity also unintendedly draws attention to, and heightens the salience of, these categorical distinctions—especially for those individuals who happen to be statistically "deviant" from the population means. In other words, the categorical outliers get noticed and are likely to feel noticed! As a result, individuals who occupy token categories may experience greater self-consciousness and heightened perception of being under evaluative scrutiny compared to nontokens. This can contribute, I argue, to a paranoid style of social perception and attribution.

To investigate the plausibility of this line of reasoning, Study 1 compared the level of paranoid cognition between tokens and their nontoken counterparts within the M.B.A. culture. Token status was defined in terms of several dimensions: gender, race, and sexual orientation. Specifically, the "token" sample included black male, black female, and "openly" gay students within the first-year M.B.A. population (N.B., "openly gay" was defined as having publicly acknowledged one's sexual orientation to classmates). The nontoken comparison group consisted of white male students.

To assess the level of paranoid social cognition within these two groups, I obtained several kinds of data. First, extensive survey data were collected on a variety of independent and dependent variables (e.g., self-consciousness, level of social discomfort, perceived status, distrust, etc.). Second, I used an event recall procedure to elicit information about group members' attributions about prior interactions with other group members that might be indicative of paranoid cognition. To disguise the purpose of the study, all of these questions were embedded within a much larger survey that was ostensibly concerned with student perceptions of the organizational culture of the M.B.A. program.

Major Results and Discussion

To assess support for the major theoretical expectations, I first examined data pertaining to group members' self perceptions and their general perceptions of their group (N.B., The differences described below were assessed using one-way analyses of variance and are significant at the $p < 0.05$ level unless otherwise indicated).

Self-Perceptions

With respect to their self-perceptions, token group members reported significantly higher levels of self-consciousness compared to their nontoken counterparts. This pattern emerged both in classroom discussions (*M*'s of 3.92 vs. 3.20 for tokens and nontokens respectively) and in social gatherings outside of class (*M*s of 3.99 vs. 3.03). This pattern was not evident, however, with respect to friends outside of the business school setting (*M*s of 1.89 vs. 2.01).

When asked how comfortable they felt discussing personal issues with other M.B.A.s and interacting with them socially (seven-point scale: 1 = Not at all comfortable to 7 = Very comfortable), tokens also reported feeling less comfortable (*M* = 3.50) compared to nontokens (*M* = 4.40). As a measure of their willingness to self-disclose to other group members, this might be regarded as a proxy for trust in the group.

Tokens also perceived themselves as fitting less well into the M.B.A. culture (*M* = 4.76) compared to nontokens (*M* = 5.67).

Perceptions of Their Group

The data pertaining to individuals' general perceptions of their groups, which can be thought of as their "stereotype" about the ingroup, also suggest a pattern consistent with the theoretical story advanced earlier. First, tokens tend to view their groups as less willing to discuss sensitive issues (*M* = 5.03) compared to nontokens (*M* = 5.90), although this difference was only marginally significant $p = .08$. They also view other M.B.A.s as less open with them personally (*M*s of 5.14 and 6.01 respectively). Tokens think their groups make it less easy to fit in compared to nontokens (*M*s of 5.40 and 6.57 respectively). And they tend to have a less positive view of their group's ability to handle conflict effectively compared to nontokens (*M*s of 4.82 and 5.33 respectively, $p = 0.08$).

In putting these patterns in perspective, it should be noted that, despite these differences, both tokens and nontokens maintain fairly positive overall stereotypes about their groups (as evidenced by the fact that, for each attribute, the absolute means even for the tokens' ratings are consistently above the midpoint of the scale). However, this can be regarded as providing particularly strong support for the theoretical predictions in the sense that the present research undertook a very conservative test of the paranoid social cognition conjecture. In other words, observing the predicted differences within a group that is generally positive and cooperative suggests strong support for the robustness of the pattern. One might expect to observe even stronger evidence in groups in which tokens stand out even more starkly or in which social differentiation introduces even higher levels of anxiety and conflict (cf. Kramer 1991).

Recall Data

The recall data were generated by asking individuals to recall and briefly describe any and all experiences they could think of that had in any way adversely affected their level of trust in either a particular group member, or the group as a whole over the past four months (note: this was the time period between orientation and the completion of their first full quarter as students). After they had generated their lists, respondents were asked to look them over and estimate for each incident how much time they had spent thinking about the incidents and /or talking about them (these time estimates were then summed for data analysis purposes).

The resulting data are interesting in several respects. First, tokens recalled significantly more instances of behaviors they felt had adversely affected their trust in the group (M = 5.72) compared to nontokens (M = 3.01). Inspection of these lists reveal that the kinds of behaviors tokens construed as violations of trust were often related to behaviors that appeared to call into question or invalidate their perceived standing in the group. For example, one student described an incident where another student "didn't bother to return a phone call to me during finals week." Similarly, another noted that a student "ignored me at a party." In many instances, a personalistic cause was proffered for this offense. For example, one student noted that, "They [sic] didn't call me to tell me about a party until it was too late and I couldn't attend because of a prior commitment. If they had wanted me to come they would have called me earlier. It was on purpose."

In addition to differential recall, there is some evidence in these data that tokens spent more time thinking about, and talking about, these incidents, compared to nontokens. For example, tokens reported spending significantly more time engaged in private (personal) ruminative activity (M = 25.43 minutes total estimated time) compared to nontokens (M = 12.12 minutes total estimated time). They also reported spending more time engaged in collective (social) rumination (M = 62.03 minutes total estimated time) compared to nontokens (M = 17.76 minutes total estimated time).

The impression of greater rumination, both privately and socially, emerges also in the descriptions of the incidents. For example, one token student reported that another (nontoken) was "really rude—she shot down my comment in class ... it was the first thing I had said all quarter. I couldn't get it out of my mind. I even talked to my friends to make sure I wasn't over-reacting ... every class after that I thought about it every time I raised my hand ."

Rater Impressions

As another way of assessing differences between tokens' and nontokens' construal of their social interactions, I had M.B.A. students from another course, and who were blind to the purpose of the study, read the recall lists and evaluate the person who made the list along a number of dimensions. I also asked these independent raters to try and form an impression of the person who had generated the list of recalled incidents. Finally, I asked them to rate the general level of "paranoia" of the person, using as a guideline the definition of paranoid cognition presented earlier in this chapter.

Several noteworthy results emerged from this procedure. First, tokens were rated as angrier than nontokens (Ms of 3.22 and 1.75 respectively), more unhappy (Ms of 2.87 and 1.66 respectively). They were also more likely to be characterized by raters as "losers" compared to nontokens (Ms of 3.74 and 1.20 respectively; seven-point scale: 1 = Not at all a loser to 7 = Very much a loser; Question, "To what extent do you think other students would characterize this person as a 'loser'?").

Perhaps most significantly, raters perceived the accounts generated by tokens as attributing more intentionality to the offenses described (Ms of 5.99 and 4.12 respectively). They also rated tokens as more paranoid (M = 3.82) compared to nontokens (M = 2.10). In interpreting this last result, it is useful to keep in mind that raters had only the recall lists from which to form an impression of the person they were evaluating (i.e., they had no individuating information about the person who had generated the list).

In putting these results in perspective, it is important to emphasize several points. First, while the data provide support for the general picture of the M.B.A. token as a relatively "paranoid" social perceiver, it is critical to note that the absolute levels of these means suggest fairly benign levels of distrust and suspicion. In general, the group of M.B.A.s used in this research consists of exceptionally talented, confident, and high self-esteem individuals. Typically, students value membership in the group and have considerable pride in membership in the M.B.A. program. Of greater relevance to the theoretical arguments, however, is the consistent pattern of *differences* between the ratings of token and nontoken group members. Second, it is necessary to emphasize that I am not claiming here that tokens' perceptions of being "over observed" are in error. Previous theory and research, in fact, would lead one to expect that, in all likelihood, the behaviors of these tokens are scrutinized more intently and that they are treated differently than other group members. In fact, in class discussions over the years, tokens often describe the M.B.A. culture as "passively" discriminating in terms of a tendency for tokens to be treated differently "at the social margins" (e.g., to not be invited to a party not because of an intentional decision not to invite, but because the thought of inviting them occurs less readily). In interpreting these data, it is also important to note that it is logically possible for individuals to both actually be disproportionately observed relative to other group members and at the same time to overestimate the extent to which they are over observed. In other words, the perceptions of tokens may contain both a "kernel of truth," and yet still reflect a *systematic* cognitive bias or distortion.

Finally, it should be emphasized that the argument is not that all individuals in token roles in organizations experience paranoid cognitions or that token status per se leads to paranoid cognition. For example, individuals whose token status is predicated upon more positive social stereotypes may view the disproportionate attention they attract as a positive feature of distinctiveness in a group. Consistent with this argument, one young female Asian M.B.A. student in the study felt she received extra "social credits" for her status. Because she was Asian and petite, she felt people assumed she was not only exceptionally bright but also modest, sincere, and hardworking. So, whether one "basks" or "bakes" in the limelight of others' attention clearly depends, at least somewhat, upon how they construe that attention.

STUDY 2: THE LIABILITY OF BEING NEW

Study 2 was designed to provide a conceptual replication of Study 1 and extend its basic findings by exploring additional determinants of paranoid social cognition in groups. Specifically, it investigates whether new members of a group are more likely to experience paranoid cognition compared to group members with longer tenure.

Newcomer Paranoia

There are several reasons why tenure within groups, defined as how long individual group members have been group members, might influence their susceptibility to paranoid cognitions. First, newcomers are likely to be more self-conscious in their social interactions compared to group members. Because they are new, they are more actively engaged in deciding how much personal information to disclose and to whom. They have to decide, for example, whether to admit such things as being new, nervous, and /or unfamiliar with organizational routines. Consequently, self-presentational concerns and impression management issues are likely to loom large during social interaction.

As newcomers, they are also likely to be undergoing a fairly intense process of socialization. For example, they must learn whatever norms exist regarding appropriate social interaction, including what is expected of them as group members, what informal social roles they will eventually occupy in the group, and the kinds of public identities that are acceptable. Newcomers often experience social anxiety during the socialization process because, although they are highly motivated to fit into and be accepted by the group, their place in its social order is still being actively negotiated. As a consequence, uncertainty and insecurity regarding their standing in the group is often relatively high. Because of uncertainty about such issues, they are likely to be relatively self-conscious in their interactions and feel a heightened sense of being under evaluative scrutiny, especially by more "senior" group members whose approval they might hope to obtain.

In contrast, those with longer tenure typically possess considerably more information regarding their standing in a group. They have already passed whatever evaluative screenings and socialization hurdles the group has placed in their path. Moreover, they are more knowledgeable about, and comfortable with, groups norms and routines. In short, they know the "ropes," and they know where they stand in the social order. As a result, uncertainty and anxiety regarding such issues are likely to be relatively low compared to newcomers. Thus, individuals with longer tenure should be less self-conscious compared to those who are new to a group.

Following this line of reasoning, I theorized that newcomers to a social group would be more self-conscious and vigilant than more experienced group members, leading them to scrutinize more intently their interactions with others for their "diagnostic" implications (e.g., signals about their standing or how well they are "fitting in").

Unfortunately, from an attributional stand point, the meaning of many interactions between group members is often uncertain or ambiguous. The cause of another group member's behavior is almost always subject to multiple interpretations, making the attribution process quite difficult (Kelley 1973). For example, the fact that one group member fails to greet another as they pass each other in the hallway at work may reflect a *personalistic* cause (the person is angry at the individual for something the individual did recently). However, there are also many *nonpersonalistic* causes that can be invoked to explain the person's behavior (e.g., he or she was preoccupied by a traffic ticket received on the way to work that morning and did not even notice the other person as they passed each other in the hallway).

From a normative stand point, an individual should discount the validity of any single causal explanation for another group member's behavior when multiple explanations for that behavior exist (Kelley 1973). Thus, even when individuals suspect they are the target or cause of another's behavior, they should discount this personalistic attribution until more conclusive evidence is available. However, despite the logical inappropriateness of doing so, there is substantial evidence that people often make overly personalistic attributions of others' actions even when competing explanations for those actions are readily available (Fenigstein 1979; Fenigstein and Vanable 1992; Greenwald 1980; Heider 1958; Kramer 1994).

In terms of their diagnostic value, interactions involving other group members are quite obviously not equally informative to newcomers. In particular, "cross-status" interactions (i.e., interactions involving group newcomers and veterans) introduce a further measure of attributional complexity to the social judgment process. For example, for the newly hired, untenured assistant professor who has recently joined a faculty department, interactions with more senior, tenured colleagues should "loom larger" than interactions involving other new assistant professors. Interactions with senior colleagues are likely to convey more diagnostic information than comparable interactions with other assistant professors regarding one's stand ing and how well one is adjusting to the group culture. Thus, assistant professors might be expected to be especially self-conscious, anxious, and vigilant when interacting with their senior colleagues compared to interactions involving other new assistant professors. Accordingly, it seems reasonable to hypothesize that the tendency to make overly personalistic attributions about others' behavior should be particularly pronounced for interactions between newcomers to a group and those with longer tenure. Relatedly, information related to such encounters should be processed more extensively in terms of post-interactional ruminative activity.

Overview of Methods

As in Study 1, I used a sample of M.B.A.s to investigate these research questions. A vignette methodology was used. Individuals were told they would be participating in a study that was concerned with how M.B.A.s evaluate their interactions with other M.B.A.s within the business school. New (first-year) and experienced (second year) M.B.A.s were asked to read a series of eight vignettes, each of which described a hypothetical interaction between the participant in the study and another M.B.A.. In each vignette, study participants found themselves in the role of "target" or "victim" of a potential violation of trust committed by either a first-year or second-year M.B.A. student. For each vignette, two attributions for the perpetrator's behavior were provided: one that suggested a personalistic (target-relevant) explanation, implying the perpetrator's behavior was intentional and directed at the target, and one that suggested a *nonpersonalistic* (target-irrelevant) explanation (a sample of the vignettes and attributions is provided in the Appendix 1). Order of presentation of the attributions was counterbalanced to control for possible order effects.

Participants were told to (1) read each vignette carefully and (2) evaluate the likelihood of the two explanations that were provided for the perpetrator's behavior. To increase involvement with the vignettes, participants were told to "really try to imagine how you would feel in this situation and how it would affect your evaluation of this person."

To summarize, the two independent variables in Study 2 were the tenure of the target or "victim" of the perpetrator's behavior (a first- or second-year M.B.A. student recruited for the study) and the tenure of the perpetrator (a hypothetical first or second year M.B.A. with whom participants imagined they were interacting). These two variables were factorially combined, producing a 2 x 2 (Target Tenure x Perpetrator Tenure) design.

Major Results and Discussion

The primary data for evaluating the attributional hypotheses are individuals' perceptions of the plausibility of the personalistic and nonpersonalistic attributions pertaining to the perpetrator's behavior. The first analysis was conducted using the personalistic attribution data alone. To facilitate analysis of these data, individuals' ratings of the likelihood of the personalistic attribution were collapsed across all eight vignettes, after first insuring that the direction of the means was the same for each vignette. This yielded a mean score reflecting each respondent's overall endorsement of the personalistic attributions.

The results of an analysis of variance of these data reveal support for both of the major hypotheses. As predicted, first-year students were significantly more likely to make personalistic attributions regarding a perpetrator's behavior (M = 3.62) compared to second-year students (M = 3.44). There was also an interaction between target and perpetrator tenure, indicating that first-year students were particularly likely to make personalistic attributions when the perpetrator was a second-year student (M = 3.92) compared to when the perpetrator was another first year student (M = 3.33). In contrast, second-year students' attributions reflected little sensitivity to the tenure of the perpetrator (Ms of 3.45 and 3.43 respectively).

Analyzing the data pertaining to respondents' personalistic attributions alone provides a direct measure of the extent to which individuals construed the perpetrator's behavior as intentional and directed at them personally. However, another way of operationalizing paranoid cognition—and one that might be viewed as a more conservative test of the hypotheses—would be to compare the extent to which individuals' make personalistic attributions *relative to* nonpersonalistic attributions. This comparative measure constitutes a more sensitive indicator of paranoid cognition in that it provides information about individuals' willingness to *discount* in (attributional terms) the plausibility of the personalistic explanation given the availability of a competing nonpersonalistic explanation. Accordingly, I also used an analysis of variance with repeated measures, with respondents' perceptions of the likelihood of the personalistic vs. nonpersonalistic attributions as a repeated factor, to evaluate the hypotheses.

This finer-grained analysis revealed a number of additional results. First, there was a highly significant effect for the repeated measure. Nonpersonalistic attributions were viewed as more likely (M = 4.68) than nonpersonalistic explanations (M = 3.53) for the perpetrator's behavior across all of the conditions. This result suggests that individuals were cognizant of the possibility of alternative explanations for the perpetrator's behavior (i.e., they weren't "completely paranoid!"). Of greater relevance to evaluating the hypotheses, there were several interactions between this repeated factor and the independent variables. First, significant interactions were observed between target tenure and the repeated measure, and perpetrator tenure and the repeated measure. Even more important, there was a significant three-way interaction involving the repeated measure, perpetrator tenure, and target tenure: first-year M.B.A. students were *most* likely to endorse the personalistic attribution (M of 3.92) and *least* likely to subscribe to the nonpersonalistic attribution (M of 4.41) when the perpetrator was a second-year student, compared to respondents in any of the other conditions. This pattern provides strong support for the hypothesis.

Table 1.   Descriptive Statistics and Correlates of Paranoid
Cognitions (Study 2)

Variable	Mean	SD	1	2	3	4
1 Paranoia	−1.14	1.09				
2 Trust in Others	5.25	1.16	−0.23***			
3 Own Trustworthiness	6.35	0.69	−0.19**	0.23***		
4 Others' Trust in You	5.29	0.87	−0.13	0.29***	0.54**	
5 Suspicion	3.62	0.80	0.56***	−0.13	0.005	−0.010

Note: *$p<0.05$; **$p<0.01$; ***$p<0.001$.

I also examined self-reports of first- and second-year student's self-consciousness to determine whether, as predicted, newcomers were more self-conscious in their interactions with other M.B.A.s compared to those with longer tenure in the organization. In support of the prediction, first-year students reported significantly higher levels of self-consciousness in their interactions ($M = 3.58$) compared to second year students ($M = 2.96$).

Confidence in the internal validity of this attribution-based analysis of paranoid cognition would be considerably enhanced if evidence were available that these attributional tendencies were correlated, in turn, with individuals' general levels of distrust and suspicion within the M.B.A. culture. Accordingly, I examined the relationship between respondents' attributional tendencies and their perceptions of collective trust, distrust, and suspicion (obtained in an earlier survey). I first constructed an individual measure of level of paranoid cognition for each respondent by creating a difference score, calculated by subtracting their nonpersonalistic attribution score from their personalistic attribution score. The relationship between this measure and their other perceptions was then assessed.

Several results are particularly noteworthy (see Table 1). First, individuals' level of paranoid cognition was highly correlated with their general level of suspicion regarding other M.B.A.s. Thus, individuals who tended to make personalistic attributions about the behavior of other M.B.A.s also tended to be more suspicious of their motives and intentions. Similarly, paranoid cognition was negatively correlated with individuals' trust in other M.B.A.s, indicating that individuals who tend to make personalistic attributions also possessed lower levels of trust in other M.B.A.s.

These results quite clearly indicate how an individual's tenure in a group can influence the attributions they make about the motives and intentions of other group members with whom they interact. In support of the theoretical arguments advanced earlier, newcomers to an organization were more likely to draw personalistic inferences from their interactions with others compared to those with longer tenure. In interpreting the theoretical significance of this pattern, it is important to note that these results do not appear to reflect a simple ingroup bias or outgroup derogation effect (cf. Brewer 1979; Cooper and Fazio 1986; Hewstone 1992; Pettigrew 1979). In particular, the interaction observed between tenure of the victim and tenure of the perpetrator suggests that the ingroup vs. outgroup status of the perpetrator alone was not driving these results (recall from Table 2 that second-year students showed little differential responsively to the perpetrator's group status).

The results of Study 2 seem particularly compelling when account is taken of the fact that the first- and second-year M.B.A. classes from which these samples were drawn were highly comparable—indeed virtually identical—with respect to "relational demographic" variables such as age, gender, educational background, and professional experience. The homogeneity of these two groups along relational demographic lines is important because it suggests that this second study provides a fairly conservative test of the paranoid attribution hypothesis. In particular, based upon a substantial body of empirical evidence (e.g., Brewer 1979; Kanter 1977b; Stephan and Stephan 1985; Tsui, Egan, and O'Reilly 1992), I would argue that one would expect to see even stronger evidence of such attributional tendencies with respect to interactions involving individuals who differ more substantially with respect to such variables (e.g., mixed-gender or cross-racial interactions). In other words, all else equal, the more "different" the interactants, the greater the propensity towards paranoid cognition one might expect to observe.

Framed more broadly, the results of Study 2 suggest some of the (cognitive) liabilities of newness in groups. Specifically, they highlight the difficulties that newcomers to a group confront when trying to get information that will help them make sense of where they stand in its social order. In this respect, these results are highly consonant with recent conceptual frameworks that characterize newcomers as highly motivated, pro-active information seekers (see, e.g., Ashford and Cummings 1985; Morrison 1993). However, they also indicate how psychological concomitants of newcomer status, such as self-consciousness and vigilance, might impede effective use of that information.

SOME CONSEQUENCES OF PARANOID COGNITION IN SOCIAL GROUPS

A number of arguments have been put forward thus far as to why categorization processes within social groups might contribute to the development of paranoid social cognition. Several cognitive mechanisms and social information processing biases have been implicated in the development of such cognitions. However, little has been said thusfar about the consequences of paranoid cognition in groups. In this section, accordingly, I describe some of the perceptual and behavioral consequences of paranoid cognition.

Paranoid Cognitions as Sources of Social Misperception

Sinister Attribution Error

The sinister attribution error, as documented above, is characterized by a tendency to over attribute hostile intentions and untrustworthy motives to others' actions. Research has shown that when individuals feel self-conscious, they often over construe others' behavior towards them in self-referential or personalistic terms. Fenigstein (1979) characterized this as the "self-as-target" bias. One consequence of the over perception of being under scrutiny by others is that it prompts scrutiny of them in return. Thus, individuals tend to "misinterpret even seemingly innocuous events as personally threatening" (Fenigstein and Vanable 1992, p. 130). This tendency to assign "sinister" meanings to other's actions is, of course, one of the hallmarks of the paranoid style of social perception. As Colby (1981) noted, "Around the central core of persecutory delusions [that preoccupy the paranoid person] there exists a number of attendant properties such as suspiciousness, hypersensitivity, hostility, fearfulness, and self-reference that lead the paranoid individual to interpret events that have nothing to do with him [sic] as bearing on him personally" (p. 518).

Biased Punctuation of Social Interaction

The biased punctuation of social interaction refers to a tendency for individuals to construe their interpersonal exchanges and transactions in a self-serving fashion (Kahn and Kramer 1990). For example, group member *A* may construe the history of distrust involving another group member *B* as a sequence *B-A, B-A, B-A* in which the initial offense or transgression in the relationship was something *B* did or did not do. As a result, *A* changes his or her behavior toward *B* in some way (e.g., acts in a less trusting manner when dealing with him or her). However, *B* punctuates the same history of interaction as *A-B, A-B, A-B,* reversing the roles of initial offender and responder. Because each person feels that the other "started it," the biased punctuation of social interaction can play an important role in the perpetuation of reciprocal cycles or "spirals" of distrust and distrust (Deustch 1973).

Exaggerated Perceptions of Conspiracy

Exaggerated perceptions of conspiracy reflect a tendency for social perceivers to overattribute coherence to a group's actions (Pruitt 1987). For example, perceptions of conspiracy in groups include perceptions of intentional snubs and slights (e.g., "They deliberately avoided telling me about the meeting or party so that I wouldn't have a chance to attend").

Behavioral Consequences of Paranoid Cognition

The impact of these cognitive processes become manifest in the kinds of defensive and avoidant behaviors that the self-conscious, paranoid individual engages in when interacting with those who make him or her feel uncomfortable. There is evidence that paranoid cognitions lead to a variety of socially maladaptive behaviors (see Kramer 1995 for a review). First, they prompt a variety of self-protective behaviors. Because paranoid individuals tend to have lower levels of trust in other group members, they tend to be more reluctant to disclose personal or sensitive information. Because of their fears regarding the extent to which others can be trusted completely, they also are more likely than others to engage in strategic disinformation to guard their true thoughts and feelings.

It is important to note that, from the perspective of the paranoid actor, such behaviors constitute wholly appropriate responses to a social environment which they perceive as hostile and fraught with psychological peril. Because they tend to entertain worst case fantasies about others' intentions and motives, leading them to underestimate others' cooperativeness and trustworthiness, they may view their own behaviors as entirely legitimate and proportioned responses to others' unfriendly, aggressive, and hostile acts.

Unfortunately, from the stand point of the targets of such behavior (who are often fairly oblivious to the paranoid actor's extreme construals), these behaviors convey the impression of a person who is distrusting, fearful, cold, and aloof—in short, unattractive and unlikable. Not surprisingly, therefore, the paranoid actor's behavior elicits the very behavior it was intended to prevent, in a self-confirming, escalatory behavioral cycle (Pruitt 1987). In this sense, the paranoid social actor and those who fuel his or her paranoia become locked into a spiral of malignant social interaction.

PULLING THE PIECES TOGETHER: A FRAMEWORK FOR CONCEPTUALIZING PARANOID SOCIAL COGNITION IN GROUPS

In the preceding pages, a variety of antecedents and consequences of paranoid social cognition have been identified. The interrelationships among these elements have not, however, been described in any detail. It may be useful, therefore, to attempt to weave the various strands of theory and evidence together into a general, integrative framework. The model shown in Figure 1 reflects an attempt to provide such a conceptual road map.

According to the model, the antecedents of paranoid social cognition include a heightened form of dysphoric self-consciousness. Dysphoric self-consciousness is an aversive state of self-awareness that centers around individuals' preoccupations that they are under evaluative scrutiny. As a result of this painful self-conscious-

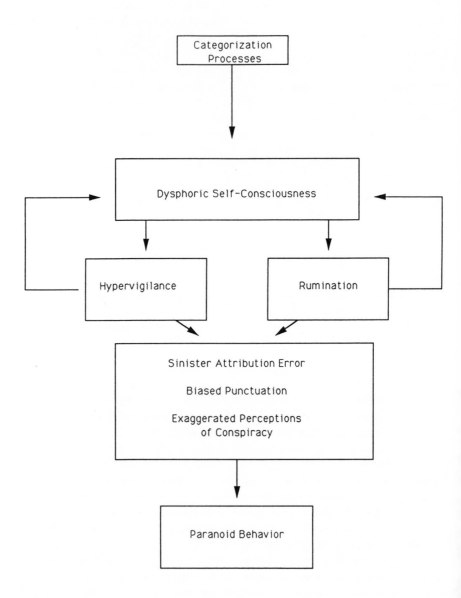

Figure 1

ness, their self-esteem and social identity may be threatened, and they may experience social anxiety and a heightened state of perceived vulnerability (cf. Fenigstein 1979; Kramer 1994, 1995).

The model further posits that this dysphoric self-consciousness fosters a hypervigilant and ruminative style of social information processing. Hypervigilance is an extreme form of vigilance associated with elevated levels of arousal, fear, anxiety, and threat (see Janis 1989; Janis and Mann 1977 for overviews). In contrast with adaptive vigilance, which can help individuals in their attempts to make sense of and cope with their social environments, there is evidence that hypervigilance contributes to social misperception and miscalculation (Janis 1989; Kramer 1995).

In some respects, it is somewhat ironic that, in response to the self-consciousness that attends individuals' feelings of being under intense scrutiny by others, they respond by engaging in a hypervigilant scrutiny of others' actions in return. This reciprocal dynamic tends to set in motion patterns of self-defeating "action-reaction" cycles, whereby others may actually end up treating the self-conscious person in ways that fuel further self-consciousness (see Kramer 1995 for a review of theory and evidence).

According to the model, a second consequence attributed to dysphoric self-consciousness is that it prompts ruminative activity. Rumination is a cognitive process that entails the excessive reliving and reanalyzing of past events, especially those that are self-relevant. Recent research on the cognitive consequences of dysphoric rumination (Kramer 1994; Lyubomirsky and Nolen-Hoeksema 1993; Pyszczynski and Greenberg 1987; Wilson and Kraft 1993) indicates several reasons why rumination might contribute to paranoid perceptions and attributions. First, it has been shown that rumination following negative events tends to increase negative thinking about those events. Additionally, there is evidence that rumination, especially rumination by mildly depressed individuals, contributes to a pessimistic attributional style. There is also evidence that the more individuals ruminate about others' behavior in competitive contexts, the greater the tendency to make "sinister" attributions regarding their intentions and motives. As with hypervigilance, there is an irony here: the individual engages in intense rumination precisely in order to help make sense of and reduce the perceived threat in their social environment (cf. Janoff-Bulman 1992). Yet, rumination only fuels and inflames their initial suspicions in a self-confirming way. Moreover, there is good experimental evidence that rumination appears to foster overconfidence when interpreting ambiguous events, leading to insufficient testing and inadequate probing of the veridicality of one's assumptions.

The impact of hypervigilance and rumination on social judgment is manifested, according to the model, in terms of the three forms of social misperception described earlier: (1) the sinister attribution error, (2) the biased punctuation of social interaction, and (3) the exaggerated perception of conspiracy.

At the behavioral level, these perceptual tendencies lead the paranoid social actor to engage in limited and ineffective reality testing. Moreover, they are likely to adopt defensive styles of interpersonal interaction, further compounding their interpersonal difficulties. While intended to protect the individual, these behaviors can be quite costly and even self-destructive (see Kramer 1995 for a further elaboration of evidence linking paranoid cognition with self-defeating social behavior).

GENERAL DISCUSSION

Earlier in this paper, I noted that the study of distrust and suspicion remains a much neglected topic in group research despite a resurgence of interest in other kinds of group phenomena. The words *distrust* and *trust* do not appear even once in the index of a number of recent and excellent edited books on groups (e.g., the volumes by Mullens and Goethals 1987; Hackman 1990). Interestingly, even when trust and distrust have been discussed in research on groups, they have often been characterized as if they were of interest primarily or only because they were mediators of other more important dependent variables, such as cooperation or conflict. A goal of this chapter has been to suggest that distrust and suspicion within social groups are important topics in their own right. Secondarily, it has tried to suggest that there are interesting dynamics of intragroup distrust and suspicion not captured by extant models of interpersonal or intergroup trust.

In putting the contributions of the present research in perspective, it may be useful to begin with a few caveats. First, the terms *paranoid social cognition* and *irrational distrust* may seem excessively pejorative labels. In effect, they seem to "blame the victim." For example, characterizing the cognitive processes of individuals who happen to occupy relatively disadvantaged positions within groups (such as newcomers or tokens) as "paranoid" might seem to minimize the legitimacy of their concerns. Relatedly, it seems to underestimate the situational origins of their plight.

This is far from the intent of the present analysis. Rather, the spirit of the analysis is to suggest that categorization processes in groups activate a variety of fairly routine and "normal" social cognitive processes that, when unchecked and under the "right" circumstances, potentially promote misconstruals and misattributions which undermine the development of interpersonal trust. A better understanding of these "irrational" bases of distrust and suspicion, I would argue, is a critical first step toward the development of more efficacious behavioral technologies for building and restoring group trust.

On similar grounds, the term *sinister attribution error* implies a mistaken or flawed process of social inference, seeming once again to cast aspersions on the cognitive competence of certain group members. In so far as the results of the present study document that psychological processes such as self-consciousness and the perception of being under evaluative scrutiny do lead to systematic distortions in the attribution process, the terms *error* and *bias* seem quite appropriate. However, it is important not to misconstrue such cognitive errors as "errors" in a more existential sense. In many groups, the risks and costs associated with misplaced trust may be quite substantial. For example, in highly competitive or political groups, a propensity towards vigilance with respect to detecting the lack of trustworthiness of other group members may be quite prudent and adaptive. In such groups, after all, it may be better for deviants to be "safe than sorry."

Such arguments prompt consideration of other possible adaptive functions that paranoid cognitions serve in group life. While the arguments up to this point have emphasized almost exclusively their deleterious consequences, there are several ways in which the psychological processes associated with paranoid cognitions (i.e., heightened vigilance and rumination) may have adaptive consequences. First, as noted above, distrust is not always irrational. Even though the fears and suspicions of group members may seem exaggerated to outsiders, this doesn't mean that their distrust is necessarily without foundation. The expression, "Just because you're paranoid doesn't mean they *aren't* out to get you," often contains more than a kernel of truth. These cognitive advantages, it should be emphasized, may be especially salient to individuals who are relatively disadvantaged with respect to their power or status in a group. In situations where the costs of misplaced trust are severe, being safe rather than sorry may be prudent policy.

When viewed from this perspective, psychological processes such as vigilance and rumination may be quite useful. In much the same way that defensive pessimism contributes to a form of *adaptive* over preparedness when individuals anticipate challenging events (Norem and Cantor 1986), so might paranoid cognitions help individuals maintain the motivation needed to overcome perceived dangers and obstacles within their social environments, even if those dangers and obstacles are—from the perspective of more objective observers—exaggerated. After all, at the very heart of the dilemma confronting the token group member or the newcomer to a group is not simply *whether* to trust or distrust but rather *how much* trust and distrust are appropriate. They must also decide who to trust and when to trust. Ultimately, the question becomes not whether distrust and suspicion are good or bad but rather, "How much is enough?" And when? And with whom? These are complex social judgments, and consequential trade-offs attend each of the possible errors in judgment (viz., misplaced trust vs. misplaced distrust) that one can make.

In addition, it is important to note that, in mild form, the cognitive processes associated with paranoid cognition are usually quite adaptive (Janis and Mann 1977). For example, vigilance and mindfulness—which might be construed as the more normal variants of hypervigilance and rumination—are often quite useful in helping individuals make sense of the social situations they are in and helping them determine appropriate forms of behavior in them.

With these caveats in mind, the present research makes a number of contributions to our understanding of the origins and dynamics of distrust and suspicion within social groups. First, although a great deal of progress over the past decade has been made in conceptualizing the antecedents and consequences of trust and distrust, an unfortunate schism remains apparent in much of the discipline-based literatures on these topics. Specifically, there has been a tendency for trust and distrust to be conceptualized primarily in terms of *either* micro or macro level processes (see Sitkin and Roth 1993 for a notable exception, however). For example, psychological research on trust has examined in great detail the role that cognitive processes such as attributions and expectations play in the development of trust and distrust (Deutsch 1973; Rotter 1967). It has also afforded considerable attention to identifying patterns of behavioral interaction that increase or decrease trust (Lindskold 1978; Rotter 1980). However, these psychological theories have remained generally "acontextual," at least in so far as they ignore the impact of important social and organizational structures. In contrast, sociological research has paid considerably more attention to the organizational structures and processes that influence trust and distrust. For example, researchers have examined how networks and other forms of social "embeddedness" influence the dynamics of trust and distrust (see, e.g., Granovetter 1985; Shapiro 1987). However, the psychological processes that mediate individuals' response to such structures and processes have remained largely unspecified.

In response to this state of affairs, a number of scholars (e.g., Barber 1983; Lewis and Weigert 1985; Sitkin and Roth 1993) have suggested the need for more integrative theory regarding trust and distrust. In particular, they have proposed that what is needed is theory and research that articulates more clearly the linkages between the micro level, psychological underpinnings of distrust and the various social contexts within which problems of distrust arise. Along these lines, several sociologists have suggested recently that more attention needs to be paid to the relational aspects of trust. Problems of trust do not arise, they have noted, in isolation from pre-existing social ties or links; rather, they are intimately embedded in ongoing social and collective relations. Accordingly, as Lewis and Weigert (1985) have argued, trust should be conceptualized as "a property of *collective units (ongoing dyads, groups, and collectivities)*, not of isolated individuals [and] psychological states taken individually" (p. 968). The present research has tried to articulate at least some of these crucial links between trust-related cognitions and the social contexts in which they are embedded.

DIRECTIONS FOR FUTURE RESEARCH

The results of the two studies described in this paper suggest several promising directions for future research. First, there has been considerable interest recently in the topic of diversity. In particular, researchers have been interested in developing a better understanding of how diversity affects social relations within groups (see, e.g., Tsui, Egan, and O'Reilly 1992; Tsui and O'Reilly 1989). The analysis presented in this chapter suggests that social heterogeneity within groups can, at least under some circumstances, contribute to a form of exaggerated distrust and suspicion, especially when social interactions are construed in categorical terms. Indeed, research on intergroup anxiety and contact (Stephan and Stephan 1985), as well as research on the effects of "social differentiation" on conflict in small groups (Kramer 1989) already provides some support for this conjecture. However, clearly, more research is needed in this area.

Another area that seems ripe for further development concerns approaches to reducing irrational distrust and suspicion within social groups. In approaching this problem, we can benefit from research on the development of trust and cooperation at the interpersonal and intergroup levels. Generally, researchers in these other domains have approached the problem of eliciting and maintaining trust in terms of either structural, behavioral, or psychological solutions. Structural solutions that have been proposed include: (1) establishing communication processes that improve the clarity of social information, thereby reducing attributional ambiguity and misperception about others' intentions and motives, (2) implementing monitoring systems for detecting violations of trust, and (3) imposition of sanctioning systems which provide reassurance that trust violators will be punished. Behavioral approaches include a variety of social influence strategies that are designed to foster the induction and maintenance of trust and cooperation (see Axelrod 1984; Bendor, Kramer, and Stout 1991; Lindskold 1978). Finally, psychological approaches include the use of cognitive strategies which foster more accurate "reality-testing" and reduce misperception in social interactions (Lord and Saenz 1985).

It might seem as if, in aggregate, the theme of the present research is rather downbeat in terms of emphasizing how easily things can go awry entirely "inside the head" of group members, leading to the erosion of trust and cooperation. However, as Lord and Saenz (1985) aptly note, "If tokens [and others] can think their way into a cognitive dilemma, they can think their way out of it too" (p. 926). The challenge for future research is to take this task to heart.

ACKNOWLEDGMENTS

The development of these ideas benefitted greatly from an interdisciplinary seminar on trust and norms organized by Jim Baron. I am grateful to Joel Podolny for providing an opportunity to collect pilot data for this research. Enjoyable discussions with Bill Barnett, Ron Burt, Jane Dutton, Jim March, Barry Markovsky, Joanne Martin, Michael Morris, Charles O'Reilly, Jeff Pfeffer, Woody Powell, Bob Sutton, Mayer Zald, and one anonymous co-editor of *Advance in Group Processes* contributed to the development of these ideas.

APPENDIX

Sample Vignette Items From Study 1

1. You call a (first/second) year student one night in the middle of finals week and leave a message on their phone that you have an urgent question regarding a final exam and could they please call you back that evening, no matter how late. You never hear from them.

How likely is it that:

a. They never received your message and /or were unavailable to call you back.
b. They heard your message but decided not to call you back.

2. You run into a (first/second) year student during midterms. You don't know the person well, but you have talked a few times at informal social gatherings in the business school. The person seems rather frantic and asks whether they can borrow some money for lunch, promising to pay you back after the exams are over. You loan them $5. Later, during the following quarter, you run into the person in the cafeteria again. They are buying a cup of coffee and have a 10 dollar bill in their hand. They are quite pleasant to you, making small talk, but never raise the issue of the loan or offer to pay you back.

How likely is it that:

a. They simply forgot about the loan.
b. They were never serious about repaying it.

3. You are having lunch with a group of students that you have just met. You are telling a joke that you consider quite funny. Suddenly, in the middle of telling the joke, one of the (first/second) year students in the group gets up and leaves.

How likely is it that:

a. They thought your joke was uninteresting.
b. They actually did have an appointment.

REFERENCES

American Psychological Association 1987. *Diagnostic and Statistical Manual of Mental Disorders.*
    Washington, DC: APA.
Ashford, S. J., and L. L. Cummings. 1985. "Pro Active Feedback Seeking: The Instrumental Use of the
    Information Environment." *Journal of Occupational Psychology* 58: 67-79.

Ashforth, B. E., and F. Mael. 1989. "Social Identity Theory and the Organization." *Academy of Management Review* 14: 20-39.

Axelrod, R. 1984. *The Evolution of Cooperation.* New York: Basic Books.

Barber, B. 1983. *The Logic and Limits of Trust.* New Brunswick, NJ: Rutgers University Press.

Baron, J. N., and J. Pfeffer. 1994. "The Social Psychology of Organizations and Inequality." *Social Psychology Quarterly,* in press.

Bendor, J., R. M. Kramer, and S. Stout. 1991. "When in Doubt: Cooperation in a Noisy Prisoner's Dilemma." *Journal of Conflict Resolution* 35: 691-719.

Bies, R. J. 1987. "The Predicament of Injustice: The Management of Moral Outrage." Pp. 289-319 in *Research in Organizational Behavior,* Vol. 9, edited by L. L. Cummings and B. M. Staw. Greenwich, CT: JAI Press.

Bies, R. J., and J. Moag. 1986. "Interactional Justice: Communication Criteria of Fairness." Pp. 43-55 in *Research on Negotiations in Organizations,* Vol. 1, edited by R. J. Lewicki, B. H. Sheppard, and M. H. Bazerman. Greenwich, CT: JAI Press.

Blake, R. R., and J. Mouton. 1986. "From Theory to Practice in Interface Problem Solving." Pp. 67-87 in *Psychology of Intergroup Relations,* 2nd ed., edited by S. Worchel and W. G. Austin. Chicago: Nelson-Hall.

Brewer, M. B. 1979. "In-Group Bias in the Minimal Intergroup Situation: A Cognitive-Motivational Analysis." *Psychological Bulletin* 86: 307-324.

_____. 1981. "Ethnocentrism and Its Role in Interpersonal Trust." In *Scientific Inquiry and the Social Sciences,* edited by M. B. Brewer and B. E. Collins. New York: Jossey-Bass.

_____. 1991. "The Social Self: On Being the Same and Different at the Same Time." *Personality and Social Psychology Bulletin* 17: 475-482.

Brewer, M. B., V. Dull, and L. Lui. 1981. "Perceptions of the Elderly: Stereotypes as Prototypes." *Journal of Personality and Social Psychology* 41: 656-670.

Brinthaupt, T. M., R. L. Morel, and J. M. Levine 1991. "Sources of Optimism Among Prospective Group Members." *Personality and Social Psychology Bulletin* 17: 36-43.

Cameron, N. 1943. "The Development of Paranoic Thinking." *Psychological Review* 50: 219-233.

Colby, K. M. 1981. "Modeling a Paranoid Mind." *The Behavioral and Brain Sciences* 4: 515-560.

Cooper, J., and R. H. Fazio. 1986. "The Formation and Persistence of Attitudes That Support Intergroup Conflict." Pp. 183-195 in *Psychology of Intergroup Relations,* 2nd ed., edited by S. Worchel and W. G. Austin. Chicago: Nelson-Hall.

Cota, A. A., and K. L. Dion. 1986. "Salience of Gender and Sex Composition of Ad Hoc Groups: An Experimental Test of Distinctiveness Theory." *Journal of Personality and Social Psychology* 50: 770-776.

Crocker, J., and R. Luhtanen. 1990. "Collective Self-Esteem and Ingroup Bias." *Journal of Personality and Social Psychology* 58: 60-67.

Deutsch, M. 1973. *The Resolution of Conflict.* New Haven, CT: Yale University Press.

Fenigstein, A. 1979. "Self-Consciousness, Self-Attention, and Social Interaction." *Journal of Personality and Social Psychology* 37: 75-86.

Fenigstein, A., and P. A. Vanable. 1992. "Paranoia and Self-Consciousness." *Journal of Personality and Social Psychology* 62: 129-138.

Festinger, L., H. W. Riecken, and S. Schachter. 1956. *When Prophecy Fails: A Social and Psychological Study of a Modern Group That Predicted the Destruction of the World.* New York: Harper and Row.

Fiske, S., and S. Taylor. 1991. *Social Cognition,* 2nd ed.. New York: Rand om House.

Fox, A. 1974. *Beyond Contract: Power and Trust Relations.* London: Faber and Faber.

Gambetta, D. 1987. *Trust: Making and Breaking Cooperative Relations.* Cambridge, MA: Oxford.

Golembiewski, R. T., and M. McConkie. 1975. "The Centrality of Interpersonal Trust in Group Processes." Pp. 131-185 in *Theories of Group Processes,* edied by C. L. Cooper. London: John Wiley and Sons.

Granovetter, M. 1985. "Economic Action and Social Structure: The Problem of Embeddedness." *American Journal of Sociology* 91: 481-510.

Greenwald, A. G. 1980. "The Totalitarian Ego: Fabrication and Revision of Personal History." *American Psychologist* 35: 603-618.

Gurtman, M. B. 1992. "Trust, Distrust, and Interpersonal Problems: A Circumplex Analysis." *Journal of Personality and Social Psychology* 62: 989-1002.

Hackman, J. R. 1990. *Groups That Work and Those That Don't.* San Francisco: Jossey-Bass.

Heider, F. 1958. *The Psychology of Interpersonal Relations.* Hillsdale, NJ: Erlbaum.

Heller, J. 1966. *Something Happened.* New York: Laurel.

Hewstone, M. 1992. "The Ultimate Attribution Error? A Review of the Literature on Intergroup Causal Attribution." *European Journal of Social Psychology* 20: 311-335.

Janis, I. L. 1983. *Groupthink,* 2nd ed.. Boston: Mifflin.

———. 1989. *Crucial Decisions.* New York: Free Press.

Janis, I. L., and L. Mann. 1977. *Decision Making.* New York: Free Press.

Janoff-Bulman, R. 1992. *Shattered Assumptions: Toward a New Psychology of Trauma.* New York: Free Press.

Kanter, R. 1977a. *Men and Women of the Corporation.* New York: Basic Books.

———. 1977b. "Some Effects of Proportions on Group Life: Skewed Sex Rations and Responses to Token Women." *American Journal of Sociology* 82: 965-990.

Kelley, H. H. 1973. "Causal Schemata and the Attribution Process." *American Psychologist* 28: 107-123.

Kramer, R. M. 1991. "Intergroup Relations and Organizational Dilemmas: The Role of Categorization Processes." Pp. 191-227 in *Research in Organizational Behavior,* Vol. 13, edited by L. L. Cummings and B. M. Staw. Greenwich, CT: JAI Press.

———. 1994. "The Sinister Attribution Error: Origins and Consequences of Collective Paranoia." *Motivation and Emotion* 18: 199-230.

———. 1995. "In Dubious Battle: Heightened Accountability, Dysphoric Cognition, and Self-Defeating Bargaining Behavior." In *Negotiation in Its Social Context,* edited by R. M. Kramer and D. M. Messick. Thousand Oaks, CA: Sage Publications.

Kramer, R. M., M. B. Brewer, and B. Hanna. 1995. "Collective Trust and Collective Action: The Decision to Trust as a Social Decision." In *Trust in Organizations,* edited by R. M. Kramer and T. R. Tyler. Thousand Oaks, CA: Sage Publications.

Lewis, J. D., and A. Weigert. 1985. "Trust as a Social Reality." *Social Forces* 63: 967-985.

Lind, E. A., and T. R. Tyler. 1988. *The Social Psychology of Procedural Justice.* New York: Plenum.

Lindskold, S. 1978. "Trust Development, the GRIT Proposal, and the Effects of Conciliatory Acts on Conflict and Cooperation." *Psychological Bulletin* 85: 772-793.

Lord, C. G., and D. S.Saenz. 1985. "Memory Deficits and Memory Surfeits: Differential Cognitive Consequences of Tokenism for Tokens and Observers." *Journal of Personality and Social Psychology* 49: 918-926.

Luhmann, N. 1979. *Trust and Power.* New York: Wiley.

Lyubomirksy, S., and S. Nolen-Hoeksema. 1993. "Self-Perpetuating Properties of Dysphoric Rumination." *Journal of Personality and Social Psychology* 65: 339-349.

Martin, J. 1993. "Inequality, Distributive Justice, and Organizational Illegitimacy." In *Social Psychology in Organizations,* edited by J. K. Murnighan. Englewood-Cliffs, NJ: Prentice Hall.

McQuire, W. J., and A. Padawer-Singer. 1976. "Trait Salience in the Spontaneous Self-Concept." *Journal of Personality and Social Psychology* 33: 743-754.

Morrison, E. W. 1993. "Newcomer Information Seeking: Exploring Types, Modes, Sources and Outcomes." *Academy of Management Journal* 36: 557-589.

Mullens, B., and G. R. Goethals. 1987. *Theories of Group Behavior.* New York: Springer-Verlag.

Norem, J. K., and N. Cantor. 1986. "Defensive Pessimism: Harnessing Anxiety as Motivation." *Journal of Personality and Social Psychology* 51: 1208-1217.

Olson, M. 1965. *The Logic of Collective Action*. Cambridge, MA: Harvard University Press.

Paulus, P. B., M. T. Dzindolet, G. Poletes, and L. M. Camacho. 1993. "Perception of Performance in Group Brainstorming: The Illusion of Group Productivity." *Personality and Social Psychology Bulletin* 19: 78-89.

Pettigrew, T. F. 1979. "The Ultimate Attribution Error: Extending Gordan Allport's Cognitive Analysis of Prejudice." *Personality and Social Psychology Bulletin* 5: 461-477.

Pfeffer, J. 1992. *Managing With Power*. Cambridge, MA: Harvard Business School Press.

Polzer, J., R. M. Kramer, and M. Neale. 1995. *Individual and Group Illusions: Effects of Reward Structure and Threat to Self-Esteem*. Unpublished manuscript.

Pruitt, D. G. 1987. "Conspiracy Theory in Conflict Escalation." Pp. 191-202 in *Changing Conceptions of Conspiracy*, edited by C. F. Graumann and S. Moscovici. New York: Springer-Verlag.

Pyszczynski, T., and J. Greenberg. 1987. "Self-Regulatory Perseveration and the Depressive Self-Focusing Style: A Self-Awareness Theory of Reactive Depression." *Psychological Bulletin* 102: 122-138.

Rotter, J. B. 1967. "A New Scale for the Measurement of Interpersonal Trust." *Journal of Personality* 35: 651-655.

————. 1980. "Interpersonal Trust, Trustworthiness, and Gullibility." *American Psychologist* 35: 1-7.

Saenz, D. S., and C. G. Lord. 1985. "Reversing Roles: A Cognitive Strategy for Undoing Memory Deficits Associated With Token Status." *Journal of Personality and Social Psychology* 56: 698-708.

Shapiro, S. P. 1987. "The Social Control of Impersonal Trust." *American Journal of Sociology* 93: 623-658.

Sitkin, S. B., and N. L. Roth. 1993. "Explaining the Limited Effectiveness of Legalistic 'Remedies' for Trust/Distrust." *Organizational Science* 4: 367-392.

Stephan, W. G., and C. W. Stephan. 1985. "Intergroup Anxiety." *Journal of Social Issues* 41: 157-175.

Taylor, S. E. 1981. "A Categorization Approach to Stereotyping." Pp. 83-114 in *Cognitive Processes in Stereotyping and Intergroup Behavior*, edited by D. L. Hamilton. Hillsdale, NJ: Erlbaum.

Tsui, A. S., T. D. Egan, and C. O'Reilly. 1992. "Being Different: Relational Demography and Organizational Attachment." *Administrative Science Quarterly* 37: 549-579.

Tsui, A. S., and C. O'Reilly. 1989. "Beyond Simple Demographic Effects: The Importance of Relational Demography in Superior-Subordinate Dyads." *Academy of Management Journal* 32: 402-423.

Turner, J. 1987. *Rediscovering the Social Group: A Self-Categorization Theory*. Oxford: Basil Blackwell.

Tyler, T. R. 1993. "The Social Psychology of Authority." Pp. 141-160 in *Social Psychology in Organizations: Advances in Theory and Practice*, edited by J. K. Murnighan. Englewood Cliffs, NJ: Prentice-Hall.

Wharton, A. S. 1992. "The Social Construction of Gender and Race in Organizations: A Social Identity and Group Mobilization Perspective." Pp. 55-84 in *Research in the Sociology of Organizations*, Vol. 1, edited by S. Bacharach. Greenwich, CT: JAI Press.

Wilson, T. D., and D. Kraft. 1993. "Why Do I Love Thee? Effects of Repeated Introspections About a Dating Relationship on Attitudes Towards the Relationship." *Personality and Social Psychology Bulletin* 19: 409-418.

Wrightsman, L. S. 1991. "Interpersonal Trust and Attitudes Toward Human Nature." Pp. 373-412 in *Measures of Personality and Psychological Attitudes*, edited by J. Robinson, P. Shaver, and L. Wrightsman. San Diego, CA: Academic Press.

Zimbardo, P. G., S. M. Andersen, and L. G. Kabat. 1981. "Induced Hearing Deficit Generates Experimental Paranoia." *Science* 212: 1529-1531.

Zucker, L. G. 1986. "Production of Trust: Institutional Sources of Economic Structure, 1840-1920."
     Pp. 53-111 in *Research in Organizational Behavior,* Vol. 8, edited by B. M. Staw and L. L.
     Cummings, Greenwich, CT: JAI Press.

MEMBER DIVERSITY AND LEADERSHIP EFFECTIVENESS:
PROCEDURAL JUSTICE, SOCIAL IDENTITY, AND GROUP DYNAMICS

Tom R. Tyler, Heather J. Smith, and Yuen J. Huo

ABSTRACT

In this chapter, we explore the conditions under which democratic authorities can bridge differences in values and interests when those differences are associated with ethnic and religious subgroups within the larger society. The relational model of authority proposes that people will defer to group authorities they identify with even when it is not in their immediate group or self-interest because they have an over-riding long-term interest in maintaining a positive social relationship with the group. People are concerned with whether or not they are treated with dignity and respect because these aspects of their experience reflect the quality of their relationship to the group and group authorities. We show, first, how relational concerns can help manage conflicts in values and interests within groups or societies, and second, we show how identification with particular groups can facilitate or hinder a relational strategy.

Advances in Group Processes, Volume 13, pages 33-66.
ISBN: 0-7623-0005-1

How is it possible for there to exist over time a just and stable society of free and equal citizens, who remain profoundly divided by reasonable religious, philosophical, and moral doctrines? ... History suggests that it rarely is.

—J. Rawls (1993, p. 4)

Our research is concerned with identifying the conditions under which democratic authorities can bridge differences in values and interests to effectively maintain internal cohesion within groups. In particular, we focus on the problematic situation in which internal differences within groups are linked to identifiable subgroups within the larger group (Azzi 1994). Such situations are especially important because research suggests that differences in values and interests can pose an unusually difficult challenge for authorities when those differences align with peoples' affiliations to ethnic or religious subgroups within the larger society (Geertz 1973; Gellner 1987; Huntington 1968). For example, people have historically been much more likely to challenge government authority when they have moral grievances which are linked to strong affiliations with religious authorities (Kelman and Hamilton 1989).

This chapter reviews and synthesizes our recent research on the exercise of authority. Most of the studies are based on questionnaires administered either through the mail or via telephone interviews. Hence, the analysis utilizes people's self-reports concerning their attitudes and behaviors.

The research presented in this chapter has both theoretical and practical signifi cance. On a theoretical level, it tests hypotheses derived from the relational model of authority (Tyler and Lind 1992). The relational model departs from the traditionally important social exchange, or public choice view of authority. That view suggests that people react to rules and authorities in self-interested ways (Bradach and Eccles 1989; Williamson 1993). In contrast, the basic proposition of the relational model (Tyler and Lind 1992) is that people defer to group authorities even when it is not in their immediate self-interest because they have an overriding long-term interest in maintaining a positive social relationship with the group.

The degree to which an instrumental or a relational explanation accurately describes people's reactions to authorities is of practical significance to the governance of a society in which there are differences in values and/or competing interests. An instrumental model suggests that conflicts of interests or of values will have a potentially divisive impact on group cohesion because people who are motivated by their self-interest will be dissatisfied if authorities are unable to give them the outcomes they seek. Hence, authorities will inevitably lose influence as they make decisions with which group members disagree. The relational model provides a more hopeful outlook by suggesting that authorities can gain acceptance for policies that are contrary to the values and interests of individuals and groups within an organization or society by implementing and using fair decision-making procedures.

There are two aspects of the relational model of authority. First, this model proposes that people evaluate their experiences through judgments about the fairness of the procedures they experience, rather than in terms of judgments about the favorability or fairness of the outcomes they receive. Rather than evaluating their experience primarily through judgments about the quality or number of assets obtained, in either absolute or relative terms, people evaluate their experiences through judgments about *how* their outcomes are determined. This does not mean, of course, that outcomes have no effect. They do. However, that influence is small (Tyler 1994b).

Second, the relational model proposes that procedural justice judgments are linked to relational aspects of experience. In particular, people are concerned with whether or not they are treated with dignity and respect, aspects of their interaction with authorities which have little to do with the nature of the outcomes received. These aspects of experience reflect the quality of the person's relationship to the group and to group authorities. Such relational judgments are distinct from judgments of process control or outcome favorability—instrumental judgments which reflect the ability of the individual to secure desired outcomes in interactions with others. Hence, on two levels, the relational model suggests that people distinguish their social concerns from concerns about the favorability of the outcomes they receive from group authorities. The relational model predicts that people react to process evaluations. Furthermore, they make those evaluations through reference to relational aspects of their experience with authorities.

The relational model predicts that people's reactions to authorities are rooted in concerns about social identity issues. People are concerned with the fairness of procedures because the quality of interpersonal treatment they receive from group authorities is an especially potent source of information that can be used for determining whether they are respected, valued members of the group (Tyler, Degoey, and Smith 1996). People are concerned with how they are viewed within their important reference groups because such information shapes their self-images (e.g., Tajfel and Turner 1986; Tyler 1994b). This perspective is supported by research showing that fair procedures, especially when fairness is defined as treatment with neutrality, dignity and respect, have been found to influence feelings of self-worth (e.g., Koper, Van Knippenberg, Bouhijs, Vermunt, and Wilke 1993; Tyler, Degoey, and Smith 1996).

Instrumental theories of human behavior, such as the individual level social exchange theory (Thibaut and Kelley 1959) and the group level realistic group conflict theory (Taylor and Moghaddam 1994), provide a grim image of the future, since they suggest that authorities are judged by their ability to meet demands for resources. In contrast, the relational model provides an optimistic alternative. According to the relational model, authorities are not judged by what they deliver but rather by the processes they use to make allocation and conflict resolution decisions. Further, it suggests that there are mechanisms through which leaders can accomodate differences in interests and values to effectively manage diversity.

In this chapter, we will explore first how relational justice can help manage diversity in values and interests within groups or societies. Second, we will show how identification with particular groups or social categories can facilitate or hinder a relational justice strategy.

DIVERSITY AND DEMOCRATIC GOVERNANCE

One problem we address in this chapter is how issues of diversity impact the effectiveness of democratic governance. The growing diversity of American society, and the increasing importance of ethnicity and race as sources of identification, raise questions about the future viability of American democratic institutions. Other societies with ethnic and/or religious diversity have resolved their governance problems in a variety of ways—the physical separation of groups (South Africa and Yugoslavia), development of autocratic central authority (the Soviet Union); and the extermination of minorities (Turkey and Iraq). The question for our democratic society is whether diversity can be managed through liberal democracy.

Demographic projections clearly show that the ethnic and cultural nature of American society is changing. In California, America's most populous and ethnically diverse state, white Americans are projected to be a minority by the year 2002 (California Department of Finance 1993). Further, in the 1990s, 45 percent of all additions to the American work force will be non-white, with 50 percent of those additions coming from Asian and Latin countries (Cox 1993; Fullerton 1987; Johnston 1991).

Perhaps more importantly, the character of American democracy is changing. There are suggestions that current and future demographic changes may present America's political, legal, and business organizations with a different set of problems than it has dealt with in the past. America has always been a nation of immigrants. However, the traditional model of civic culture has involved "minority assimilation" of new citizens into an identification with American society and its political and social values (Moghaddam and Solliday 1991). Immigrants were expected to forsake their own cultural values and adopt, if not embrace, the Western political, legal, and social values of mainstream American society. To facilitate this goal, society provided an open educational system through which such assimilation could occur. One of the most important of the values into which people were socialized is the belief in and respect for democratic processes and individual rights.

Increasingly, however, the members of various ethnic and national subcultures are reluctant or unwilling to abandon their unique cultural values and identities to assimilate into mainstream culture. One example of this change is the strong second language movement in California. Instead of seeking to learn and teach their children English, minority citizens are pressuring for education in their own languages. In the political arena, there are pressures toward a multilingual ballot, while, in the legal system, it has been proposed that rights be created entitling people to bilingual lawyers and translators.

Such pressures for cultural heritage retention are moving America in the direction of a multicultural or mosaic society, in which distinct subgroups coexist within the framework of a single society.[1] Under this model, instead of one set of dominant social, political, and legal values, society must accommodate in some way to the values of the various subgroups within it (Lieberson and Waters 1987). This changing social structure has provoked widespread concern about the presumed destructive consequences of moving from a society based on identification with a single set of superordinate values and institutions to a society composed of citizens with strong identifications with their ethnic and racial subgroups—a "mosaic" society. Concerns about the transformation of America into a "preservative of diverse alien identities," that is, a society of "groups more or less ineradicable in their ethnic character," have been expressed by social commentators (Schlesinger 1992). Underlying these concerns is the question of whether democratic institutions and values can continue to exist in a mosaic society. The fear is that the legitimacy of superordinate group rules and authorities would diminish among minorities who place greater importance on their affiliation with their own ethnic subgroups than the superordinate group.[2]

Since it is widely recognized that America is becoming a diverse society, it is important to identify and understand the impact of that social change on the future effectiveness of our political and legal institutions. In this chapter, we present both theoretical and empirical evidence to develop the proposition that there are mechanisms through which authorities can accomodate different values and interests. Specifically, we suggest that a relational justice strategy which emphasizes people's concerns about procedural justice, relational issues, and social identification can enhance the legitimacy of authorities, and be used successfully to manage a diverse population.

LEGITIMACY AND EFFECTIVENESS

Before considering how diversity impacts leadership effectiveness, we first need to consider what general factors underlie such effectiveness. The function of authorities is to manage internal conflicts and to organize group tasks (Messick et al. 1983; Rutte and Wilke 1984; Samuelson 1991; Sato 1987; Thibaut and Faucheux 1965; Wit 1989; Yamagishi 1986a, 1986b, 1988) by bridging differences within groups to create policies which are followed by all group members. Compliance with such policies could potentially be gained in a variety of ways, including through coercion and bribery, via persuasion or expertise, or by appeals to legitimacy (French and Raven 1959). The key to authorities' effectiveness lies in their ability to secure compliance from group members.

Studies of authority suggest that the ability to gain not just forced compliance, but voluntary acceptance of rules and decisions, is especially valuable (Easton 1965; Gamson 1968; Kelman 1969, Parsons 1963, 1967; Sarat 1977; Scheingold 1974; Tyler 1990). Authorities gain such acceptance by appealing to feelings of an internalized obligation to obey, that is, "legitimacy." Legitimacy has three components: the willingness to accept the decisions of authorities, the willingness to follow group rules, and favorable evaluations of authorities (Tyler 1994c). In this chapter we will not differentiate among these aspects of legitimacy. For a more detailed discussion of these distinct aspects of legitimacy, refer to Tyler (1994c).

Authorities typically have, at best, a limited ability to compel obedience and must depend on the widespread existence of feelings that authorities are legitimate and ought to be obeyed. This is particularly true of democratic societies, such as the United States, which depend heavily upon voluntary cooperation.

Voluntary cooperation is valuable since those authorities who can effectively appeal to internal values within group members do not have to (1) expend group resources for rewards (2) create plausible threats through which to threaten group members, and/or (3) persuade and justify every decision. Instead, their decisions are voluntarily accepted because group members feel that those legitimate leaders are entitled to be obeyed. As a consequence, authorities can more effectively pursue long-term goals. It is for this reason that legitimacy has been suggested by both political theorists (Dahl 1971) and social scientists (Gamson 1968) to be key to the ability of democratic authorities to function effectively (Tyler 1990). A crucial determinant of the consequences of increasing diversity in society for the viability of democratic political systems is the impact of such diversity on the legitimacy of superordinate authorities.

PROCEDURAL JUSTICE AND LEGITIMACY

Since, as we have suggested, legitimacy is crucial for leadership effectiveness in democratic societies, it is necessary to determine the antecedents of legitimacy. One set of antecedents is suggested by the instrumental model of authority relations. The instrumental model links evaluations of authorities to judgments of direct and indirect control over the outcomes of allocation procedures (Thibaut and Walker 1975). An instrumental explanation suggests that people are linked to others through the resources they exchange and that self-interest is fundamental to such exchanges (Tyler 1994c). Therefore, judgments of legitimacy should be related to outcome favorability or policy agreement.

A second set of antecedents of legitimacy is suggested by the relational model of authority. According to the relational model of authority, legitimacy is shaped by (1) judgments about the fairness of decision-making procedures and (2) relational indicators of neutrality, trustworthiness, and status recognition. These two aspects of the procedural model are linked since neutrality, trustworthiness, and status recognition are the primary antecedents of procedural justice judgments (Tyler 1994b; Tyler and Lind 1992). The neutrality of decision-making procedures, the trustworthiness of authorities, and the quality of the interpersonal treatment which people receive from authorities are labeled relational concerns because they communicate information about the nature of the social relationship between people and the groups to which they belong (Tyler 1989; Tyler and Lind 1992).

The relationship among the antecedents of legitimacy are represented in Figure 1. This examination of the antecedents of legitimacy considers the two stages of the relational justice model. First, the model compares the direct influence of instrumental judgments, relational judgments, distributive justice, and procedural justice on legitimacy. Second, the model compares the influence of instrumental and relational judgments on definitions of the fairness of procedures and of outcomes.

The analysis presented in Figure 1 is based upon six data sets, each of which uses interviews about people's interactions with group authorities (total n = 2,298). In the analysis both aspects of the relational model receive support. First, it is procedural justice and relational judgments, and not distributive justice or instrumental judgments which are significantly related to whether people view authorities as legitimate. Further, it is relational judgments, not instrumental judgments that dominate perceptions of procedural justice. This pattern of results occurs irrespective of whether the authorities involved are legal, political, managerial, educational, or familial (Tyler 1994c). The finding that people focus on procedural justice and the finding that they define procedural justice relationally both support the effectiveness of using a relational justice strategy to bridge across competing interests. In contrast, if people focus on winning, or define fair procedures as procedures that gave them favorable outcomes, then a procedural strategy would not be effective in bridging across differences in values and interests.

PROCEDURAL JUSTICE AND POLICY ENDORSEMENT

The findings from the six data sets shown in Figure 1 are generated from interviews with individuals about their reactions to *direct* experiences with authorities. In many other situations, however, experiences with authorities are infrequent or indirect. For example, few citizens have directly interacted with representatives of the U.S. Supreme Court or other national level institutions. Furthermore, institutions like the U.S. Congress or the U.S. Supreme Court articulate general policy principles that could potentially impact many group members, not just those who are immediately involved in a particular dispute or decision. Therefore, it is important to determine whether the procedural justice effects found in the studies summarized in Figure 1 can be generalized to studies of national level institutions. This question is especially important because many of the most pervasive and per-

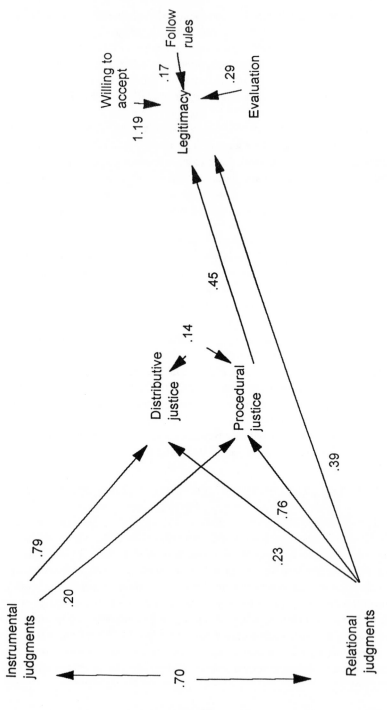

Figure 1. The Antecedents of Legitimacy

40

Table 1. Decision Making and the Legitimacy of the Supreme Court: The Case of Abortion

	Willingness to empower the court to make abortion policy	Feeling an obligation to obey government decisions	Regarding the court as a legitimate legal authority
Agreement with court decisions	0.16*	0.03	0.04
Court decisions are fair	0.26*	−0.03	0.08
The court makes decisions in a fair way	0.43*	0.31*	0.13*
The morality of abortion	0.08	−0.16*	0.11
Should abortion be legal?	−0.11	−0.02	−0.17*
Do you agree with court abortion decisions?	0.05	−0.13*	−0.1
R-squared	62%*	18%*	13%*

Note: Unless otherwise indicated, all entries are beta weights. Starred entries are significant ($p < 0.05$). Demographics controls were also included in the equation. The influence of ideology, party identification, sex, income, education, age, race, and religion were controlled for.

sistent conflicts in a multicultural society arise out of differences in opinions about important social and political issues that must be resolved at the national level. Ultimately, the viability of a diverse society depends on whether national level authorities can manage differences in policy preferences.

Recent studies suggest that national-level authorities can use procedures to effectively manage policy differences. Tyler and Mitchell (1994) explored the antecedents of people's willingness to defer to the U.S. Supreme Court on the issue of abortion rights. They did so in telephone interviews with a random sample of 502 citizens in the San Francisco Bay area. The results are displayed in Table 1. Their findings indicate that the primary antecedents of both legitimacy and the willingness to defer to the court are evaluations of the fairness of court decision-making procedures. These three judgments—legitimacy, willingness to empower, and obligation—are themselves interrelated, suggesting a general procedural influence on reactions to authorities.

Further, procedural justice judgments are relationally based (see Table 2). In a direct comparison of the influence of control over outcomes to the effect of neutrality, trustworthiness, and status recognition, the relational indicators were found to be more central to judgments about procedural justice than were judgments of control.

Table 2.   Elements of Procedural Fairness: The U.S. Supreme Court

	Procedural justice	Legitimacy	Empowerment	Obligation to obey the law
Do you have influence over court decisions?	−0.13	−0.07	0.02	0.05
Is the court neutral?	0.48*	0.63*	0.17*	0.18*
Is the court trustworthy?	0.34*	0.13*	0.22*	−0.03
Are you treated with respect?	0.10	0.07	−0.02	0.07
R-squared	60%*	56%*	12%*	5%*

Note: Unless otherwise indicated, entries are beta weights. Starred entries are statistically significant.

The findings from the Tyler and Mitchell study suggest that relationally fair procedures can facilitate policy endorsement even when people disagree with the policies. Specifically, people were willing to endorse a judicial decision derived using fair procedures even when the decision conflicts with their personal beliefs and values. The implication of this set of findings is that a relational justice strategy could be potentially very effective in reconciling individual differences in beliefs and values.

The Tyler and Mitchell study demonstrated that a relational strategy can allow authorities to effectively manage differences in religious and moral values. In this chapter, we suggest that one current challenge for democratic institutions is the increased importance of ethnic group memberships. Will a relational justice strategy still be effective when public policies directly address issues that are related to ethnic or racial memberships? In other words, can a relational justice strategy be used to facilitate endorsement of a policy which (1) has salient costs and benefits for the individual and (2) pits the interests of one group against another. A consideration of such a policy is important for two reasons. First, it examines whether fair procedures can accomodate strong differences in group-based interests. If a policy poses significant personal costs or benefits to people, then perhaps the effects of a procedural strategy would be limited. Second, it has been suggested that public policies which are associated with group membership issues may be especially divisive (Taylor and Moghaddam 1994).

Preferential treatment and anti-discrimination policies are two public policies which have great cost and benefit implications for individuals. For disadvantaged minority members, these policies are in their interests, while for advantaged majority members, these policies are not in their interests. Furthermore, such policies are particularly volatile because they are linked to group membership. Individuals are either helped or harmed by such policies because of their ascribed group membership. Using survey interviews with a random sample of people in the San Francisco area, Lea, Smith, and Tyler (1995) examined whether fair procedures can enhance endorsement of such controversial policies.

As shown in Table 3, respondents' race, their perception of their group's outcomes and their beliefs that they or their group would benefit or lose from the policy were all significantly related to their support for race-targeted policies such as preferential treatment and anti-discrimination policies. However, evaluations of policy fairness were more strongly related to support for anti-discrimination and preferential treatment policies than were (1) assessments of their likely impact on individuals or groups, (2) respondent characteristics (including race and sex), or (3) assessments of the magnitude of ethnic inequality. Hence, even in the case of controversial group-based policies which help some and not others, a procedural strategy is effective in enhancing policy support.

Furthermore, general relational evaluations of whether Congress was the appropriate and legitimate decision maker were more closely related to policy endorsement than specific evaluations of the policy's costs and benefits, regardless of the respondents' group membership. For both minority respondents, who stood to benefit from these policies, and for majority respondents, who might potentially lose, general evaluations of Congress were more closely related to policy endorsement than were assessments of the policy's costs and benefits. Again, people are willing to support authorities that they view as legitimate even if they, or their group, might not directly benefit from the policies of that authority in the short term.

Both of the studies on public support for controversial policies are surveys in which people are interviewed about their perceptions of the actions of national authorities. The conclusions of these studies were based on correlational analyses. However, one could argue that it is beliefs that the authority is legitimate that lead to judgments that procedures are fair rather than vice versa. Further support was found for the effectiveness of using a procedural strategy in a study which used experimental vignettes (Tyler 1994a). In that study, people's support for federally-funded abortions was assessed after they read one of several vignettes in which the nature of the outcome and the fairness of the procedure used by Congress to adopt the policy was systematically manipulated. The fairness of the procedure was manipulated in two ways. In half of the vignettes, someone representing the respondents' opinion had the opportunity to explain their opposition or support for the policy (voice), while in the other half of the vignettes there was no such opportunity (no voice). In half the vignettes the decision was the product of public hearings (neutral), while in the other half of the vignettes, the decision was made in private (not neutral). As shown in Table 4, both the outcome and the procedure had a significant impact on people's endorsement of a leader who supports a policy to fund abortions. Most importantly, this experiment shows that process issues such as neutrality and voice had a direct causal effect on support for a policy independent of one's personal views on the matter.

Table 3.   Antedecents of Support for Public Policies: All Respondents

	Support for anti-discrimination laws		Support for preferential treatment policies	
Character of the respondent	12%		14%	
Race		3%		9%
Age, sex, education, ideology, racial attitudes		9%		5%
Magnitude of the problem	8%		14%	
Unfair differences in outcomes		5%		7%
Unfair differences in opportunities		6%		12%
Views about the policy and policy maker	30%		37%	
The policy helps/ hurts me and my group		3%		7%
Congress is legitimate and appropriate decision maker		12%		15%
The policy is fair		27%		15%
Total R-squared	37%			44%

Note: All entries are the square of the adjusted multiple correlation coefficient. For example, an entry of 9% indicates that the characteristics of the person being interviewed explain 9% of the variance of policy support.

In summary, the existing evidence suggests that a procedural strategy can effectively manage conflicts of values and interest in both interpersonal interactions *and* in public disagreements about important social policy issues. Authorities benefit when those within their group accord them legitimacy based on judgments about the fairness of their decision-making strategies and on perceptions of their neutrality, trustworthiness, and willingness to respect group members (i.e., the antecedents of procedural justice). Leaders benefit because they are able to secure voluntary compliance with their decisions, which is more efficient than relying on their ability to govern based on reward or coercion. This procedural/relational underpinning of legitimacy is crucial to the effectiveness of authorities since it allows authorities to act without having to compel obedience to their decisions and

Table 4.   Experimental Vignette: Support for Federally-Funded Abortions

Legitimacy (Would you vote for the authority involved?)

Variable	Degree of freedom	Mean square	F-value
Outcome (A)	1	302.70	79.78***
Neutrality (B)	1	25.18	6.64***
Voice (C)	1	35.97	9.48***
A*B	1	1.08	0.29
A*C	1	5.13	1.35
B*C	1	9.80	2.58
A*B*C	1	0.11	0.03
Evaluation of Congress	1	18.37	4.84*
Views of the issue	1	2.29	0.60
Age	1	4.07	1.07
Gender	1	1.13	0.30
Race	1	0.03	0.01
Income	1	0.42	0.11
Education	1	0.51	0.13
Residual	432	3.79	
Total	446	4.63	

Note: Favorable or unfavorable outcomes, neutral or biased decision-making procedures, and presence or absence of an opportunity to present opinions were manipulated in the experimental vignette. Other variables were measured.

to social rules. Legitimacy gives authorities discretionary power which can be used to serve the long-term interests of the group.

In contrast, if legitimacy were based on policy agreement or outcome favorability, then authorities would lose legitimacy whenever they created policies which countered the self-interests or values of individuals or groups within society. However, research studies have failed to support the instrumental model of authority. Instead, they have supported the predictions of the relational model which provides a hopeful conclusion about the ability of authorities to bridge across differences among subgroups within a larger social context. It suggests that authorities can garner wide spread support for their policies, as well as make difficult decisions without losing public support, if they utilize fair decision-making procedures and show respect for individual members of their society.

Existing research shows that the procedural justice approach has considerable ability to help authorities accomodate among differences in values and interests about important social problems. Such an approach allows authorities to effectively manage intractable differences in opinion about what the "right" solution to these problems should be. For example, there are many solutions to the current crisis in health care. The research suggests that if people perceive that the policy adopted to deal with the health care crisis is arrived at using fair procedures, then there is a higher likelihood that they would accept the policy than if the procedures used are perceived as unfair, even if they might benefit from an unfair policy. This is the key advantage of a relational justice strategy.

However, we must be wary of possible abuses of a procedural strategy. Fair procedures can be enacted to distract individuals from their true interests. Research has demonstrated how people's attention can be diverted away from distal, distributive issues to proximal, procedural issues. In fact, people may use the fairness or unfairness of procedures as a heuristic for determining the justice of the outcome distribution (Folger 1987; Lind, Kulik, Ambrose, and De Vera Park 1993). Perceptions of a procedure as fair might discourage a critical assessment of outcomes. The heuristic that fair procedures lead to fair outcomes (Tyler and Lind 1992) might be overused, leading to relative satisfaction with otherwise undesirable results (e.g., Greenberg 1990; Tyler and McGraw 1986). Scheingold (1974), for example, has examined the use of courts by members of disadvantaged groups and suggests that the courts often provide only symbolic satisfactions. That is, the disadvantaged derive satisfaction because they focus on the proximal procedural issues involved in litigation, rather than on the distal distributive problems that they initially hoped to solve. Focusing on the fairness of procedures can obscure the fact that nothing has changed.

DIVERSITY AND PROCEDURAL CONSENSUS

One possible limitation of the use of procedures is that they require a consensus about what the elements of a fair procedure are. If people of varying ethnicities, ages, gender, or social classes do not agree on the criteria which define a fair procedure, then it is not feasible to implement procedural strategies in diverse environments. In such an event, everyone would agree that they would defer to a fairly reached decision, but authorities may find it difficult to create a procedure that all parties would judge to be fair. In this situation, the relational model would be effective in theory but impossible to implement in practice.

Several studies suggest that ethnicity and other demographic characteristics have very little influence on how people evaluate the fairness of procedures. In a study of hypothetical interpersonal dispute situations, Lind, Huo, and Tyler (1994) examined preferences for dispute resolution procedures among four ethnic groups in the United States—European Americans, African Americans, Asian Americans, and Hispanic Americans. They found that there was a remarkable degree of consensus about people's preferences for procedures. More importantly, the findings show that procedural justice concerns were more important than either outcome favorability or concerns about reducing conflict in determining preferences for procedures. Demographic characteristics such as ethnicity and gender did not

Table 5.   Demographic Influences on the Importance of Different Criteria in Defining the Meaning of the Procedural Justice of Congress

	Race	Gender	Educ.	Income	Age	Ideol.
Importance of outcome favorability (OF)	0.15*	0.16**	0.04	0.03	0.05	0.12*
Importane of voice (V)	0.02	0.00	0.06	0.09	0.00	0.08
Importance of neutrality (N)	0.39***	0.25***	0.36***	0.38***	0.32***	0.28***
Importance of trust (T)	0.18*	0.31***	0.19***	0.22***	0.23***	0.30***
Importance of respect (R)	0.08	0.13*	0.18**	0.10	0.17**	0.12
Main effect of demographic (race, gender, etc.) (DEMO)	−0.06	−0.06	−0.04	−0.03	−0.01	0.01
DEMO * OF	−0.07	−0.07	0.10	0.10	0.08	−0.03
DEMO * V	0.01	0.03	−0.05	−0.07	0.05	−0.07
DEMO * N	−0.10	0.06	−0.09	−0.10	−0.03	0.03
DEMO * T	0.07	−0.10	0.09	0.02	0.01	−0.08
DEMO * R	0.07	0.03	−0.07	0.07	−0.03	0.03
R-squared	44%	44%	44%	44%	43%	43%

Note: The dependent/criterion measure is procedural justice. Each column represents the particular demographic characteristic (e.g., race) included in the equation as both a main effect and as part of the interaction terms

affect the relative importance of procedural justice concerns in determining procedural preferences.

Not only are people's preferences for particular procedures essentially the same, so are the criteria that they use for determining whether procedures are fair. Tyler (1988) examined the influence of demographic characteristics on judgments about the fairness of personal experiences with the police and courts. He found that people who varied in their age, gender, race, education, and income level used the same criteria to make procedural justice judgments. Hence, the demographic characteristics of the person involved in a dispute did not affect perceptions of procedural justice. Interestingly, the nature of the issue or problem involved was important. People did not see the same procedure as fair in resolving all problems. Hence, the crucial issue shaping views about procedural fairness was the nature of the problem, not the character of the person.

Tyler (1994a) conducted a similar analysis but in a context where people's reactions are less likely to reflect their personal experiences. Using interviews with a random sample of citizens in the San Francisco area he examined people's evaluations of the fairness of Congressional decision-making procedures and found no differences in the criteria used to evaluate procedural justice which were linked to the demographic characteristics of the respondents (see Table 5). None of the demographic characteristics tested, race, gender, education, income, age, or ideology, interacted with relational evaluations in predicting procedural justice judgments. Similar findings were found in a study of people who were interviewed about past real disputes (Lind, Tyler, and Huo 1995). The findings of that study showed that there were only minor differences in how Americans, Germans, and Hong Kong Chinese evaluated the fairness of procedures.

Interestingly, a recent report on the future of the California courts identified the lack of consensus about how people want the courts to resolve their disputes as one of the major problems which increased diversity poses for the legal system (Dockson 1993). The research reported here suggests that this concern with a lack of procedural consensus is overstated. Cross-cultural research does suggest that procedural justice principles do vary (e.g., Leung and Lind 1986), but the widest and most influential cross-cultural differences occur for principles of distributive and retributive justice (e.g., Hamilton and Sanders 1992; Miller and Bersoff 1992; Murphy-Berman, Berman, Singh, Pachuri, and Kumar 1984). In contrast, people seem to have relatively similar views about what constitutes a fair procedure for resolving a problem or dispute.

Should we be concerned about such large differences in people's views about distributive and retributive justice? The procedural justice approach suggests that, while such differences are real, they may not be central to the effectiveness of authorities. People will defer to authorities if those authorities follow fair procedures. Hence, it is procedural justice rather than distributive or retributive justice that is central to the effectiveness of authorities.

THE PSYCHOLOGICAL UNDERPINNINGS OF RELATIONAL EFFECTS

The findings presented suggest that when people evaluate authorities, they consider more than just their potential instrumental rewards and costs. People also consider how fairly authorities treat group members. Hence, authorities are able to gain acceptance for their decisions based on the neutrality of their decision-making procedures, their trustworthiness, and their respect for those within their group. In other words, they have a non-instrumental basis for exercising authority and are thus able to handle differences in interests and values.

A key premise of the relational model is that people's concerns about the information communicated by fair procedures and respectful, dignified treatment are linked to issues of social identification. As suggested by social identity theory, the relational model assumes that people use groups as a source of information about themselves (Hogg and Abrams 1988, 1990; Tajfel 1978; Tajfel and Turner 1986).

As salient and prototypical group representatives, group authorities communicate information about people's value to the larger group through their actions (Hogg and Abrams 1988; Tyler and Lind 1992). Fair and respectful treatment accords importance and status to individual group members while unfair and disrespectful treatment communicates marginality.

This analysis is supported by the finding that people care about the relational aspects of the actions of authorities. In particular, people focus on the quality of interpersonal treatment by authorities (Tyler 1994c; Tyler and Lind 1992). While neutrality and trustworthiness may plausibly be viewed as having both identity and resource antecedents, it is difficult to explain concerns about treatment with respect and dignity from a resource perspective. On the other hand, such concerns seem quite plausibly related to people's efforts to use their treatment by authorities as information about their self-worth.

As would be predicted by the relational model, people's treatment by authorities has been found to have an important impact on their evaluations of their value to the group and, through it, their self-esteem (Koper, Van Knippenberg, Bouhuijs, Vermunt, and Wilke 1993; Tyler, Degoey, and Smith 1995; Vermunt, Wit, Van den Bos, and Lind 1993). If people were interested in their personal gain, their self-esteem would be enhanced by winning. In fact, however, self-esteem is enhanced by receiving fair treatment.

The linkage between fair interpersonal treatment and social identification issues suggests that there may be limits to people's concerns about relational concerns. Whether authorities treat group members in a neutral, trustworthy, and respectful way should only matter if the authorities are included within important reference groups. If the authorities are not part of an important group, what they think will have little impact, if any, on individuals, and hence, relational issues should be relatively unimportant.

In an experimental study, the affiliation of an authority was experimentally manipulated to represent either an ingroup (a member of the same university as the subjects) or an outgroup (a member of a rival university). Perceptions that they had been unfairly treated only influenced students' self-esteem when the authority represented an ingroup, not when the authority represented an outgroup (Ortiz 1994). These results support the argument that relational and procedural concerns should be more important when people identify with the group.

SOCIAL CATEGORIES AND PROCEDURAL JUSTICE

In addition to implications about how people are affected by their treatment by others, the relational model also has implications for how people treat others. The importance of group memberships for defining behavior toward others is illustrated by research on the "scope" of justice (Deutsch 1985; Opotow 1990). The scope of justice describes how group memberships define the relevant boundaries within which justice and relational concerns are important (Deutsch 1985; Opotow 1990; Tyler and Lind 1990). For those who fall outside of a person's scope of justice, issues of fair play will be less relevant (Deutsch 1985; Opotow 1990).

The consequences of being excluded from the scope of justice are illustrated in a study which showed that people have higher standards of justice for ingroup members (a social/political group they belong to) than for outgroup members (a so cial/political group they dislike) (Huo 1994). While people consistently granted community resources, procedural rights, and equal and fair treatment to ingroup members, they were less willing to do so for outgroup members (see Figure 2). Further, the most consistent predictor of willingness to exclude others from considerations of justice is identification with the target group. This finding supports the suggestion that justice concerns are linked with identification processes.

The pattern of results displayed in Figure 2 also indicates that people are more likely to deny resources to a disliked other than to deny them fair and polite treatment. This pattern provides further support for the relational model. It suggests that people feel that the worst thing that can be done to someone is to deny them decent treatment as a human being. On the other hand, denying a person resources seems less serious.

Even though people were the least willing to exclude outgroup members on the basis of fair treatment, they still responded in a way that shows they think that it is more important for ingroup members than for outgroup members to receive fair treatment. This finding demonstrates the important role that social categorization or group membership plays in determining when justice matters. While people's willingness to exclude outgroup members depends upon whether they are denying resources, rights, or treatment in general, the findings suggests that justice is more important in interactions with other ingroup members than in interactions with outgroup members.

This research suggests that group boundaries are important to authority relations in two ways. First, authorities can be included or excluded from particular "moral communities." If the authority is included within important reference groups, their behaviors and policies communicate relational information. Furthermore, their treatment of any single group member can communicate, by inference, their attitudes and opinions toward all group members (suggesting why the more indirect and abstract policy decisions of national-level institutions can communicate relational information). However, if particular authorities are excluded from important reference groups, relational aspects of their behavior should be irrelevant. Similarly, other people can be included or excluded from particular "moral communities." If people view the targets of particular public policies as outside of their scope of justice, the relational fairness of the procedures used to reach policy decisions should not influence their opinions. How these "outsiders" are treated by authorities does not have any implications for how the authorities view them. In contrast, if the targets are included within their scope of justice, how the targets are

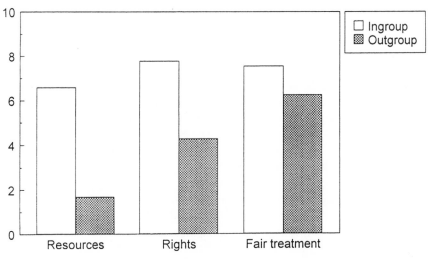

Figure 2. Entitlement to Resources, Rights, and Fair Treatment

treated communicates relational information that can be important for all group members.

Recent research supports the hypothesis that relational evaluations should be most important in interactions with ingroup members. The findings show that when a conflict crosses group boundaries (e.g., cultural boundaries), people define fairness in more instrumental terms (Tyler, Lind, Ohbuchi, Sugawara, and Huo 1995). That study explored conflicts between and among Western and Japanese teachers. Each subject, whether Western or Japanese, described a conflict which occurred (1) within their own cultural group and (2) with a member of the other cultural group. When the conflict occurred with someone outside of the subject's group, subjects both acted and evaluated their experience in more instrumental terms. In contrast, within group conflicts are evaluated in terms of relational concerns. Similarly, Tyler, Lind, and Huo (1995) examined the exercise of authority within and across ethnic categories. In that study, they investigated workers' experiences in resolving a dispute with their supervisors. They found that people evaluated authorities more strongly in terms of the fairness of the procedures they used and less strongly in terms of instrumental issues when those authorities were from their own ethnic group.

While the research findings generally support the relational model of authority, there are situations under which people become less concerned with procedural justice and relational issues. The same people evaluate authorities differently depending upon whether they categorize those authorities as being within or outside of important reference groups. Ingroup authorities are evaluated mainly on the basis of relational issues whereas outgroup authorities are evaluated more on their ability to provide favorable outcomes.

This research outlines an important implication for understanding the impact of diversity in a democratic society. On the one hand, people can include members of different ethnic groups within a larger shared collective category. In this case, we would expect relational principles of justice to be more important than instrumental evaluations, as occurred when people described a conflict with a person who shared the same cultural background. On the other hand, people may view members of ethnic groups that are different from themselves as outgroup members. In this case, we would expect instrumental evaluations to be more important, as occurred when people described a conflict with a person who did not share the same cultural background. The relational model suggests that the effectiveness of democratic leaders should be facilitated by common identification with American society; superordinate identification should facilitate solutions to the problems of diversity within the larger society. However, as identification with different subgroups in a society takes on greater importance, intergroup relations within the superordinate group can take a destructive turn if people view members of other groups as competitors in a world where not everyone's needs can be met.[3]

SUPERORDINATE IDENTIFICATION AND AUTHORITY EVALUATIONS

The key question is how the operation of procedural justice is influenced by issues of identification. The relational model suggests that identification with the category the authority represents should facilitate the influence of relational judgments on perceptions of legitimacy and policy acceptance—attitudes and behaviors that benefit the group. Social dilemma research illustrates how shared social categorizations can encourage group-oriented behavior (Kramer 1991; Wit 1989). When a common group membership was made salient to participants, the number of collectively-oriented cooperative responses to social dilemmas compared to individually-oriented competitive responses increased dramatically (Brewer and Kramer 1986; Wit 1989). When a common group membership is made salient, people will prefer to maximize their group's outcomes even if it might decrease their personal benefits (Brewer and Kramer 1986; Markovsky 1985).

Investigations of ingroup bias illustrate the same principle but focus on the relationship between groups and a larger superordinate categorization (Gaertner, Dovidio, Anastasio, Bachman, and Rust 1993; Gaertner, Mann, Dovidio, Murrell, and Pomere 1990). High school students who described themselves as American, in addition to a racial or ethnic identity, were less likely to discriminate between their own and other ethnic groups than students who only mentioned their ethnic identity (Gaertner, Rust, Dovidio, Bachman, and Anastasio 1994). When students identify themselves as members of a category that includes diverse ethnic subgroups, they are less likely to discriminate. This research supports our suggestion that identification with a superordinate category can facilitate a relational orientation toward potential outgroup members.

Although this research illustrates how identification with relevant groups can facilitate behavior that benefits the group, it has not explored how the functioning of procedural justice is shaped by the degree to which people identify with the superordinate group. Identification with a more inclusive social category should not only decrease ingroup bias, it should also change how people define procedural justice. One effort to examine how identification influences the dynamics of reactions toward authorities focused on a social dilemma situation in which societies were dealing with scarce resources. That study examined people's willingness to defer to and empower authorities to deal with a scarcity problem.

Tyler and Degoey (1995) considered a naturally occurring example of such a situation—the California water shortage. They interviewed a random sample of 402 residents of San Francisco in an effort to explore the antecedents of judgments about the legitimacy of the water commission—the agency which makes rules of water allocation. They compared the influence of a combined index of procedural and relational concerns to the influence of judgments about the favorability of the decisions of the commission. Their basic findings support the conclusions of prior procedural justice research. They found that procedural/relational judgments about the actions of authorities influence their legitimacy. People are more willing to support authorities and their rules if those authorities act in a procedurally fair way. Further, they are more willing to give power to those authorities. The key additional analysis conducted by Tyler and Degoey is an exploration of the relationship between identification with the overall community (community pride) and the role of procedural/relational concerns in legitimizing authorities. The results in Table 6 suggest that those who identify with their community evaluate authorities in more strongly relational terms.

The finding that superordinate identification is linked with a relational focus in evaluations of superordinate authorities is also supported by two other studies. Smith and Tyler (1996) studied a sample of 352 white residents in the San Francisco Bay area who were interviewed concerning their attitudes about public policies against workplace discrimination and in favor of redistributive economic programs ("affirmative action"). This study explored the willingness of the advantaged to support congressional policies which benefit the disadvantaged. Respondents were asked to evaluate the fairness of congressional decision-making procedures, and their evaluations were linked to their willingness to support congressional policies. The study also assessed identification with the superordinate community—in this case "Americans." As shown in Table 7, the study found that, in general, a combined index of procedural justice judgments and relational concerns had a strong relationship to policy endorsement. Those who felt that Congress made decisions fairly were more willing to support government policies, irrespective of whether those policies favored them.

Table 6.   Support for Government Authority

	The legitimacy of authority	The empowerment of authority
Main effects		
Instrumental judgments about authorities	0.06	0.35***
Relational judgments about authorities	0.43***	0.22***
Interactions		
Connection with the community (pride) and instrumental judgments	−0.22**	−0.08
Connection with the community (pride) and relational judgments	0.26**	0.21**
Main effect of connection with the community (pride)	0.03	−0.02
R-squared	25%***	22%***

Further, as Tyler and Degoey found, in forming judgments about policy support, those who identified more strongly with the superordinate community relied more heavily on procedural/relational judgments, and less heavily on instrumental/distributive judgments. If respondents did not identify strongly with the overall American society, they decided whether to support congressional policies by evaluating whether those policies benefited them. However, if they identified strongly with American society, they decided whether to support congressional policies by evaluating whether Congress made decisions fairly (i.e., in neutral ways, with benevolent motives, and with respect for citizen rights). Again, leaders benefited from people's identification with the superordinate group.

In a study of organizational employees, Huo, Smith, Tyler, and Lind (1996) also showed that employees' reactions to how their supervisors handled a work conflict were determined by the degree to which they identified with the work organization (see Table 8). People who identified highly with the work organization reacted to how their supervisors handled their conflict in terms of whether they were fairly treated. In contrast, people who identified less with the work organization evaluated their supervisors in terms of whether they received favorable outcomes.

The findings presented illustrate how superordinate identification can enhance the effectiveness of authorities. Authorities benefit from people's identification with the group or category they represent, since those who identify highly with the group evaluate authorities in terms of relational issues such as their neutrality and trustworthiness, as well as through their treatment of citizens. Those who do not identify with the group focus more strongly on whether the decisions of the

Table 7. Superordinate Identification and Policy Endorsement

	Policy endorsement
Identification with Americans (*IDENT*)	0.09
Instrumental evaluations (*I*)	0.20***
Relational evaluations (*R*)	0.33***
*IDENT * I*	0.15**
*IDENT * R*	−0.10*
R-squared	17%***

Note: Unless otherwise noted, the entries are standardized regression coefficients.

Table 8. Superordinate Identification and Decision Acceptance

	Reactions to authorities
Instrumental judgments (*I*)	0.96**
Relational judgments (*R*)	0.07
Superordinate identification (*IDENT*)	−0.05
*IDENT * I*	−0.54**
*IDENT * R*	0.53**
R-squared	86%

Note: Unless otherwise noted, the entries are standardized regression coefficients.

authorities favor them. Superordinate identification enhanced the legitimacy of authorities, and hence, their ability to elicit voluntary compliance with decisions which were not congruent with the short-term interests of some group members.

SUBGROUP IDENTIFICATION AND AUTHORITY RELATIONS

Using superordinate identification to facilitate a relational focus in interactions with authorities provides a fairly optimistic outlook for bridging across competing interests among subgroups within a larger group. However, identification with the group represented by a particular authority is just one of many different possible groups or social categories with which people might identify (Gaertner et al. 1993; Turner, Hogg, Oakes, Reicher, and Wetherell 1987). People might also identify with smaller or subordinate groups that are included within the larger or superordinate group. In an increasingly diverse society such as America, there is a strong possibility that, for some people, identification with an ethnic subgroup might supersede identification with America. This section of the chapter deals with the question of whether the ability of superordinate group identification to facilitate a relational focus in evaluations of authorities is limited when people have strong attachments to subgroups within the larger group.

Past research has identified two categories of people that are of particular interest to the study of competing identities. The first category includes people who draw a greater sense of identification from the superordinate group and its authorities and institutions than a particular subgroup. In essence, these people represent assimilators—they have forsaken subgroup loyalties and values in favor of the superordinate group. The second category includes people who draw a greater sense of identification from a subgroup than from the superordinate category. These people represent separatists—they have chosen to identify more closely with the subgroup and its values than with the superordinate category.

The distinction between assimilators and separatists suggests how group identification can be a double-edged sword. On the one hand, if people identify strongly with the superordinate category (America) which includes all the subgroups, they will focus on relational issues when evaluating authorities. In contrast, people who identify more with their subgroup than the superordinate group may draw less of their sense of self from the larger society. Hence, they may be indifferent to the relational messages communicated by superordinate authorities as well as authorities who represent other subgroups, focusing instead on the outcomes they can obtain.

In addition to assimilators and separatists, there is also a third category important to discussions of identification. This third group, the biculturalists, has been recently recognized in the ethnic identification literature and consists of those who identify strongly with both their superordinate category and their subgroup (Berry 1984; LaFromboise, Coleman, and Gerton 1993). A more comprehensive picture of the influence of identification can be obtained by examining all three groups: the assimilators, the biculturalists, and the separatists. As shown in Figure 3, a consideration of both subgroup and superordinate identification suggests four different categories.[4] Three groups, assimilators, separatists, and biculturalists, are included in the following analyses. The fourth possibility, the alienated (i.e., those who do not identify with either the subgroup or the superordinate group), was dropped from the analyses. The relational model of authority assumes that people are motivated to belong to groups and does not provide theoretical predictions for individuals who attach little value to group memberships.

The biculturalists pose an interesting test of the effects of multiple levels of group identification. There are three possibilities. The biculturalists could react as the separatists would and, hence, pose problems for outgroup authorities who are unable to provide them with favorable outcomes. In contrast, they could react as the assimilators would and evaluate authorities in terms of relational issues rather than the authorities' ability to meet their demands. A third possibility is that the psychological underpinnings of their reactions to authorities would not be dominated by either relational or instrumental concerns but rather jointly and equally determined by both.

Two studies have explored the dynamics of procedural justice among the assimilators, the biculturalists, and separatists. They examined the extent to which the

	Superordinate identification	
Subgroup identification	High	Low
High	Biculturalist	Separatist
Low	Assimilated	Alientated

Note: Adapted from Berry (1984).

Figure 3. Superordinate and Subgroup Identification

ability of authorities to function effectively is compromised by (1) the existence of dual identities (biculturalists) and (2) the dominance of subgroup identity (separatists). In the study of white respondents' attitudes toward affirmative action, both identification with the superordinate category (Americans in general) and with a subordinate category (racial group) was measured (Smith and Tyler 1996). Three distinct groups were identified: assimilators, biculturalists, and separatists. Assimilators identified more strongly with "Americans in general" than with their own racial group. Biculturalists identified strongly with both groups. Finally, separatists identified most strongly with their own racial group.

An examination of the basis of policy endorsement within each group suggests that procedural/relational concerns dominated policy endorsement among the assimilators (people identify primarily with America) and the biculturalists (people who equally identify with America and with their ethnic group) but not among separatists (people who identify primarily with their ethnic group) (see Figure 4). Among separatists, policy support was dominated by instrumental judgments. The findings of the Smith and Tyler study suggest that the procedural justice approach to bridging differences may have a clear limit. It is much less successful among separatists. On the other hand, the findings suggest that assimilation, and hence the relinquishing of subgroup identification, is not necessary for a procedural strategy to be effective. Assimilation and biculturalism lead to a very similar psychological basis for reactions to group authorities—a focus on procedural and relational issues.

One limitation of the Smith and Tyler study is that it examines abstract policy preferences. What about the willingness to accept the decisions of authorities in concrete everyday interactions? A further limitation is the focus on a single racial group—white Americans. This latter limitation is particularly problematic since one of the primary concerns about increasing diversity is the extent to which ethnic minorities are more loyal to their subgroup than to the superordinate group. Hence, it is important to examine whether a relational strategy would work among members of ethnic minority groups. To address these limitations, we investigated workers' evaluations of their experience in conflicts with their work supervisor. A deliberate effort was made to recruit respondents from a wide range of ethnic backgrounds. In that study, Huo, Smith, Tyler, and Lind (1996) interviewed 305 employees of the University of California at Berkeley. The employees were

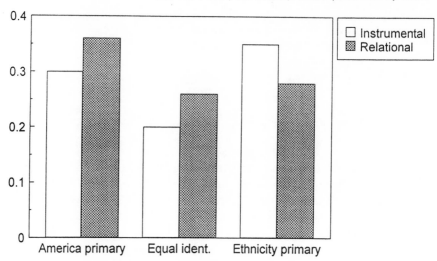

Figure 4. Superordinate Identification, Subgroup Identification, and Policy
Endorsement

recruited through ethnic employee associations which represented a variety of ethnic groups including European Americans, Asian Americans, African Americans, and Hispanic Americans. Each respondent was interviewed about a recent experience with a supervisor about an important work related issue. In this analysis, procedural and relational concerns were distinguished, and each was compared to an index of outcome favorability.

Like Smith and Tyler (1996), this study introduced issues of subgroup identification. Employees were divided into three groups: the assimilators, the biculturalists, and separatists. As Figure 5 shows, analysis within each subgroup indicated that the assimilators and those who are bicultural both rely on relational judgments in reacting to decisions made by their supervisors. Separatists, however, put greater weight on the instrumental index. The findings show that superordinate identification is essential to using a relational strategy. Further, identification with a subgroup does not pose a problem as long as there is also a social bond with the superordinate group.

In an era of increasing cultural diversity, ethnic identification is a natural domain for the examination of the effects of competing social identifications. Hence, it has been the focus of our recent studies. However, there is no theoretical reason that the logic of the subgroup analysis we conducted needs to be limited to issues of ethnic identification. Religious affiliation could be another potent source of subgroup identification in a pluralistic society. In fact, historically identification with religious leaders has been a major source of difficulty for government authorities (Kelman and Hamilton 1989). In such situations the moral principles articulated by religion conflict with the legitimacy of the state.

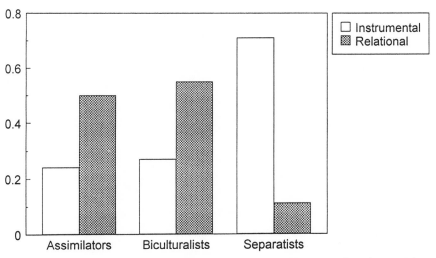

Figure 5. Decision Acceptance Among Assimilators, Biculturalists, and Separatists

Studies of why people obey the law suggest that two different sources influence people's feelings of obligation to obey the law: morality and the legitimacy of law and legal authority (Tyler 1990). Religious beliefs are one important source of morality. Hence, morality can be a source for resisting decisions or requests by government authorities otherwise viewed as legitimate (Kelman and Hamilton 1989). This suggests that religious identification may shape justice concerns in the same way that ethnic identification does.

If we return to the earlier survey of people's willingness to defer to the U.S. Supreme Court on the issue of abortion rights (Tyler and Mitchell 1994), we can illustrate our hypothesis. First, as a proxy for superordinate identification, we can distinguish between respondents who viewed the court as very legitimate and respondents who viewed the court as illegitimate. For the sake of our argument, we will assume that attachment to the superordinate group, America, and viewing its authorities as legitimate are psychologically similar. Second, we can use people's view that abortion is not moral as a proxy for religious identification. Although thinking abortion is not moral is not the same thing as religious identification, the findings indicate that those people who indicate a religious affiliation are significantly more likely to say that abortion is not moral than are those people who report no religious affiliation. As shown in Figure 6, the same pattern found in studies of ethnic identification emerges for religious identification. For "assimilators" (those who feel the court is legitimate and believe that abortion is moral) and "biculturalists" (those who feel the court is legitimate and believe that abortion is immoral), relational judgments were more closely related to feelings of obligation

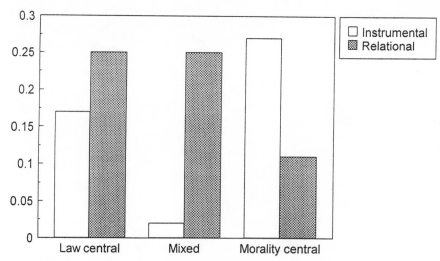

Figure 6. Superordinate Identification, Subgroup Identification, and the Court: The Antecedents of Feeling Obligated to Defer to the Decisions of Federal Authorities

to obey Supreme Court decisions. But for "separatists" (those who feel the court is illegitimate and believe that abortion is immoral), instrumental judgments were more closely related to feelings of obligation.

Although this analysis is speculative and the operationalization of relevant constructs does not map perfectly onto our hypotheses, it does illustrate how the same dynamics can be extended to include identification with a variety of different types of groups. The findings presented above also illustrate the important function legitimacy serves in eliciting feelings of obligation to defer to the decisions of authorities.

CONCLUSIONS

The question addressed in this chapter is whether there are strategies which group authorities can adopt to manage internal diversity. Two such mechanisms were identified in the research outlined. The first is that there is general support for the argument that legitimacy is linked to overall evaluations of procedural justice and to aspects of decision making which shape procedural justice judgments (neutrality, trustworthiness, and status recognition). A procedural focus stemming from the relational model of authority is the key to effectively managing diversity. Second, the ability of authorities to function in a relational way is enhanced when people have a strong social bond with the superordinate group they represent.

The findings also suggest that, in order for authorities to function effectively, people do not have to abandon their subgroup identities. Those who assimilate and those who are bicultural appear to share a similar psychology; their evaluations of authorities stem mainly from concerns about their relationship to the group, rather than concerns about the favorability of their outcomes. However, separatists may pose a problem for authorities. This group bases their evaluations of authorities on short-term instrumental concerns (i.e., what outcomes does the authority give them). Consequently, unless separatists are given the outcomes they seek, they will evaluate authorities negatively and resist complying with the decisions authorities make.

Because superordinate identification is so important to the effectiveness of a relational strategy, the question of how to enhance the quality of the social bond people have with the superordinate group should be addressed. One way to enhance identification with the superordinate group is to encourage children to appreciate and adopt the values of the superordinate group. This socialization process can be achieved through public institutions such as the education system. A second way is to encourage the learning of a common language. While the merits of a bilingual education are sources of much debate, they should not negatively impact societal cohesion for children to learn a subgroup's language, as long as children also learn the dominant language. In other words, it is more important that children do learn English than that they do not learn a second language. A third way is to show that groups included within the larger category (as well as individual members) are valued and worthy of respect (Thompson, Kray, and Lind 1994).

However, this strategy, coined "balanced multiculturalism" by Moghaddam and Solliday (1991), may be the most tricky to pursue. On the one hand, perhaps the best way to engender feelings of personal respect and trust is to demonstrate respect and trust for one's group. On the other hand, an increased emphasis on subgroup memberships may inadvertently communicate that one's ethnic group membership is more important (and therefore, a more valuable source of information about one's self-worth) than one's national group membership. Furthermore, the increased salience of particular group memberships may begin the invidious cycle of intergroup comparisons and competition (Tyler, Huo, and Smith 1995).

The research reviewed presents some interesting findings about the effects of biculturalism—identifying strongly with both the superordinate group and a subgroup. If the effective exercise of authority requires assimilation, as is often presumed in public discussions of the possible balkanization effects of multiculturalism, then the traditional "melting pot" model is important. It may be, as Rawls provocatively suggests, that it is difficult for a society to survive without a single set of values. On the other hand, if the bicultural model is supported, there is a substantially stronger possibility that a democratic society can flourish when people maintain loyalty to subgroups as long as they also identify with the larger society. Further, it has been suggested that bicultural identity also has positive psychological effects for the individual (LaFromboise, Coleman, and Gerton 1993). Ethnic subgroup identities are particularly important to people's sense of self and, consequently, difficult to suppress (Gurin and Epps 1975). Our findings from the preliminary studies outlined suggest that as long as a strong sense of superordinate identity is maintained, subgroup identities need not be relinquished in order for authorities to effectively manage relations among subgroups. Hence, it appears that future discussions of ways to manage a diverse society should be more open to issues related to subgroup identity.

Our research also has important implications for issues of democratization. As the nature of American society changes toward a mosaic model, the question of whether and how democratic processes can be maintained becomes central to discussions of public policy. While much has been written about the probable effects of ethnic and racial subgrouping on American society, and there are many examples of the political and social difficulties which such subgrouping can pose in recent events in Yugoslavia, the former Soviet Union, and Lebanon, there is very little empirical research on the effects of subgroup identifications on the functioning of stable democracies such as the United States. The research reported in this chapter examined those effects upon two central judgments—the legitimacy of authorities and policy acceptance. While focused on this issue, the research is intended to more broadly examine how the functioning of democratic authority is influenced by the development of subgroup identifications.

Bridging across differences in values and interests that are linked to identifiable subgroups is a difficult, but not impossible, challenge for superordinate authorities. Our research suggests that if authorities use relationally fair procedures, they can effectively manage internal diversity. However, there may be limits to the effectiveness of the relational justice strategy. On the one hand, people who identify strongly with the superordinate category will place more importance on relational concerns and, consequently, will be more likely to accept outcomes or policies that are not in their personal or group interests. On the other hand, people who identify less strongly with the superordinate category will place more importance on instrumental concerns. If the outcomes or policies are inconsistent with their personal or group interests, they will be less likely to accept them. However, this problem only occurs when people identify *more* strongly with their subgroup than with the superordinate category. People who identify equally with both their subgroup and the superordinate category still rely on relational evaluations of authorities. In summary, this research suggests how we can avoid the destructive consequences of ethnic divisions within society without sacrificing ethnic diversity.

ACKNOWLEDGMENT

This paper is based on a presentation to the Society of Experimental Social Psychology, Incline Village, Nevada, October 1994. We thank Leigh Thompson for organizing that symposium.

NOTES

1. An interesting macrolevel question which is beyond the scope of this paper is why there are differences between the willingness of earlier and contemporary immigrants to assimilate into mainstream values.
2. Although this concern is currently manifesting itself in the context of ethnicity there is a long history of similar concerns in the arena of church-state relations. Historically, a major source of subgroup identification was religion (Kelman and Hamilton 1989).
3. This research also illustrates how important it is to distinguish psychological identification from demographic characteristics. Demographic characteristics such as gender or ethnicity are not necessarily good predictors of attitudes and behaviors. Demographics only shape attitudes and behaviors if they are central to people's sense of themselves (Koch 1993).
4. Figure 3 is based on the premise that identification with a particular subgroup and a particular superordinate category can represent two separate orthogonal continuums (Azzi 1994; Berry 1984; LaFromboise, Coleman, and Gerton 1993). For example, strong identification with a superordinate category does not require weak identification with the subgroup.

REFERENCES

Azzi, A. E. 1994. "From Competitive Interests, Perceived Injustice, and Identity Needs to Collective Action: Psychological Mechanisms in Ethnic Nationalism." In *Nationalism, Ethnicity, and Violence*, edited by B. Kapferer. Oxford, UK: Oxford University Press.

Berry, J. W. 1984. "Cultural Relations in Plural Societies: Alternatives to Segregation and Their Sociopsychological Implications." Pp. 11-27 in *Groups in Contact*, edited by N. Miller and M. Brewer. San Diego: Academic Press.

Bradach, J. L., and R. G. Eccles. 1989. "Price, Authority and Trust." *Annual Review of Sociology* 15: 97-188.

Brewer, M. B., and R. M. Kramer. 1986. "Choice Behavior in Social Dilemmas: Effects of Social Identity, Group Size and Decision Framing." *Journal of Personality and Social Psychology* 50: 543-549.

California Department of Finance. 1993. *Projected Total Population of California Counties: 1990 to 2040*. Sacramento, CA: California Department of Finance.

Cox, T. 1993. *Cultural Diversity in Organizations*. San Francisco: Berrett-Koehler.

Dahl, R. 1971. *Polyarchy*. New Haven, CT: Yale University Press.

Deutsch, M. 1985. *Distributive Justice*. New Haven, CT: Yale University Press.

Dockson, R. R. 1993. *Justice in the Balance: Report of the Commission on the Future of the California Court*. San Francisco: Supreme Court of California.

Easton, D. 1965. *A Systems Analysis of Political Life*. Chicago: University of Chicago Press.

Folger, R. 1987. "Reformulating the Preconditions of Resentments: A Referent Cognitions Model." In *Social Comparison, Social Justice, and Relative Deprivation*, edited by J. Masters and W. Smith. Hillsdale, NJ: Lawrence Erlbaum Associates.

French, J. R. P., Jr., and B. Raven. 1959. "The Bases of Social Power." In *Studies in Social Power*, edited by D. Cartwright. Ann Arbor, MI: Institute for Social Research.

Fullerton, H. N. 1987. "Labor Force Projections: 1986-2000." *Monthly Labor Review*, 19-29.

Gaertner, S. L., J. F. Dovidio, P. A. Anastasio, B. A. Bachman, and M. C. Rust. 1993. "The Common Ingroup Identity Model: Recategorization and the Reduction of Intergroup Bias." In *The European Review of Social Psychology*, Vol. 4, edited by W. Stroebe and M. Hewstone. New York: John Wiley.

Gaertner, S. L., J. A. Mann, J. F. Dovidio, A. J. Murrell, and M. Pomare. 1990. "How Does Cooperation Reduce Intergroup Bias?" *Journal of Personality and Social Psychology* 59: 692-704.

Gaertner, S. L., M. C. Rust, J. F. Dovidio, B. A. Bachman, and P. A. Anastasio. 1994. "The Contact Hypothesis: The Role of a Common Ingroup Identity on Reducing Intergroup Bias." *Small Group Research* 22: 267-277.

Gamson, W. 1968. *Power and Discontent*. Homewood, IL: Dorsey.

Geertz, C. 1973. "The Integrative Revolution: Primordial Sentiments and Civil Politics in the New States." In *The Interpretation of Cultures*, edited by C. Geertz. New York: Basic Books.

Gellner, E. 1987. *Culture, Identity, and Politics*. Cambridge, UK: Cambridge University Press.

Greenberg, J. 1990. "Looking Fair Versus Being Fair: Managing Impressions of Organizational Justice." *Research in Organizational Behavior* 12: 111-157.

Gurin, P., and E. Epps. 1975. *Black Consciousness, Identity and Achievement*. New York: Wiley.

Hamilton, V. L., and J. Sanders. 1992. *Everyday Justice: Responsibility and the Individual in Japan and the United States*. New Haven, CT: Yale University Press.

Hogg, M. A., and D. Abrams. 1988. *Social Identifications: A Social Psychology of Intergroup Relations and Group Processes*. London and New York: Routledge.

_____. 1990. "Social Motivation, Self-Esteem, and Social Identity." In *Social Identity Theory: Constructive and Critical Advances*, edited by D. Abrams and M. A. Hogg. New York: Springer-Verlag.

Huntington, S. P. 1968. *Political Order in Changing Societies*. New Haven, CT: Yale University Press.

Huo, Y. J. 1994. *Are There Limits to Justice? Exclusionary Effects in Justice Behavior*. Unpublished master's thesis. Berkeley: University of California.

Huo, Y. J., H. J. Smith, T. R. Tyler, and E. A. Lind. 1996. "Superordinate Identification, Subgroup Identification, and Justice Concerns: Is Separatism the Problem; Is Assimilation the Answer? *Psychological Science* 7: 40-45.

Johnston, W. 1991. "Global Work Force 2000: The New World Labor Market." *Harvard Business Review* 69: 115-127.

Kelman, H. C. 1969. "Patterns of National Involvement in the National System: A Sociopsychological Analysis of Political Legitimacy." In *International Politics and Foreign Policy*, edited by J. Rosenau. New York: Free Press.

Kelman, H. C., and L. Hamilton. 1989. *Obedience to Authority*. New Haven, CT: Yale University Press.

Koch, J. W. 1993. "Is Group Membership a Prerequisite for Group Identification?" *Political Behavior* 15: 49-60.

Koper, G., D. Knippenberg, F. Bouhuijs, R. Vermunt, and H. Wilke. 1993. "Procedural Fairness and Self-Esteem." *European Journal of Social Psychology* 21: 630-650.

Kramer, R. M. 1991. "Intergroup Relations and Organizational Dilemmas: The Role of Categorization Processes." *Research in Organizational Behavior* 13: 191-228.

LaFromboise, T., H. L. K. Coleman, and J. Gerton 1993. "The Psychological Impact of Biculturalism: Evidence and Theory." *Psychological Bulletin* 114: 395-412.

Lea, J. A., H. J. Smith, and T. R. Tyler. 1995. "Predicting Support for Compensatory Polices: Why, Why, and How." Unpublished manuscript. Berkeley: University of California.

Leung, K., and E. A. Lind. 1986. "Procedural Justice and Culture: Effects of Culture, Gender and Investigator Status on Procedural Preferences." *Journal of Personality and Social Psychology* 50: 1134-1140.

Lieberson, S., and M. C. Waters. 1987. "The Location of Racial and Ethnic Groups in the United States." *Sociological Forum* 2: 780-810.

Lind, E. A., Y. J. Huo, and T. R. Tyler. 1994. "...And Justice for All: Ethnicity, Gender and Preferences for Dispute Resolution Procedures." *Law and Human Behavior* 18: 269-289.

Lind, E. A., C. T. Kulik, M. Ambrose, and M. V. de Vera Park. 1993. "Individual and Corporate Dispute Resolution: Using Procedural Fairness as a Decision Heuristic." *Administrative Science Quarterly* 38: 224-251.

Lind, E. A., T. R. Tyler, and Y. J. Huo. 1995. *Procedural Context and Culture: Variations in the Antecedents of Procedural Justice Judgments*. Unpublished manuscript. Chicago: American Bar Foundation.

Markovsky, B. 1985. "Multilevel Justice Theory." *American Sociological Review* 50: 822-839.

Messick, D. M., H. Wilke, M. B. Brewer, R. M. Kramer, P. E. Zemke, and L. Lui. 1983. "Individual Adaptations and Structural Change as Solutions to Social Dilemmas." *Journal of Personality and Social Psychology* 44: 294-309.

Miller, J. G., and D. M. Bersoff. 1992. "Culture and Moral Judgment: How Are Conflicts Between Justice and Interpersonal Responsibilities Resolved?" *Journal of Personality and Social Psychology* 62: 541-554.

Moghaddam, F. M., and E. A. Solliday. 1991. "'Balanced Multiculturalism' and the Challenge of Peaceful Coexistence in Pluralistic Societies." *Psychology and Developing Societies* 3: 51-72.

Murphy-Berman, V., J. J. Berman, P. Singh, A. Pachauri, and P. Kumar. 1984. "Factors Affecting Allocation to Needy and Meritorious Recipients: A Cross-Cultural Comparisons." *Journal of Personality and Social Psychology* 46: 1267-1272.

Opotow, S. 1990. "Deterring Moral Exclusion." *Journal of Social Issues* 46: 173-182.

Ortiz, D. J. 1994. *When Do Outcomes Matter? Group Membership, Self-Esteem, and Justice Evaluations.* Honors thesis. Berkeley: University of California.

Parsons, T. 1963. "On the Concept of Influence." *Public Opinion Quarterly* 27: 63-82.

_____. 1967. "Some Reflections on the Place of Force in Social Processes." In *Sociological Theory and Modern Society*, edited by T. Parsons. New York: Free Press.

Rawls, J. 1993. *Political Liberalism.* New York: Columbia.

Rutte, C. G., H. Wilke, and D. M. Messick. 1987. "Scarcity or Abundance Caused by People or the Environment as Determinants of Behavior in the Resource Dilemma." *Journal of Experimental Social Psychology* 23: 208-214.

Samuelson, C. D. 1991. "Perceived Task Difficulty, Causal Attributions, and Preferences for Structural Change in Resource Dilemmas." *Personality and Social Psychology Bulletin* 17: 181-187.

Sarat, A. 1975. "Support for the Legal System." *American Politics Quarterly* 3: 3-24.

Sato, K. 1987. "Distribution of the Cost of Maintaining Common Resources." *Journal of Experimental Social Psychology* 23: 19-31.

Scheingold, S. A. 1974. *The Politics of Rights.* New Haven, CT: Yale University Press.

Schlesinger, A. M., Jr. 1992. *The Disuniting of America: Reflections on a Multicultural Society.* New York: Norton.

Smith, H. J., and T. R. Tyler. In press. "Justice and Power: When Will Justice Concerns Encourage the Advantaged to Support Policies Which Redistribute Economic Resources and the Disadvantaged to Willingly Obey the Law?" *European Journal of Social Psychology.*

Tajfel, H. 1982. *Human Groups and Social Categories.* New York: Cambridge University Press.

Tajfel, H., and J. Turner. 1986. "The Social Identity Theory of Intergroup Behavior." In *Psychology of Intergroup Relations*, edited by S. Worchel. Chicago: Nelson Hall.

Taylor, D. M., and F. M. Moghaddam. 1994. *Theories of Intergroup Relations.* New York: Praeger.

Thibaut, J., and C. Faucheux. 1965. "The Development of Contractual Norms in a Bargaining Situation Under Two Types of Stress." *Journal of Experimental Social Psychology* 1: 89-102.

Thibaut, J., and H. H. Kelley. 1959. *The Social Psychology of Groups.* New York: Wiley.

Thibaut, J., and L. Walker. 1975. *Procedural Justice: A Psychological Analysis.* Hillsdale, NJ: Lawrence Erlbaum Associates.

Thompson, L., L. Kray, and E. A. Lind. 1994. *The Bright and Dark Side of Group Identity.* Paper presented at the Annual Society for Experimental Social Psychology Convention, Lake Tahoe, CA.

Turner, J. C., M. A. Hogg, P. J. Oakes, S. Reicher, and M. S. Wetherell. 1987. *Rediscovering The Social Group: A Self-Categorization Theory.* Oxford, UK: Blackwell.

Tyler, T. R. 1988. What is Procedural Justice? Criteria Used by Citizens to Assess the Fairness of Legal Procedures. *Law and Society Review* 22: 301-355.

_____. 1989. "The Psychology of Procedural Justice: A Test of the Group Value Model." *Journal of Personality and Social Psychology* 57: 830-838.

_____. 1990. *Why People Obey the Law.* New Haven, CT: Yale University Press.

———. 1994a. "Governing Amid Diversity: The Effect of Fair Decision-Making Procedures on the Legitimacy of Government." *Law and Society Review* 28: 809-831.

———. 1994b. "Psychological Models of the Justice Motive." *Journal of Personality and Social Psychology* 67: 850-863.

———. 1994c. *The Psychology of Legitimacy.* Unpublished manuscript. University of California, Berkeley.

Tyler, T. R., and P. Degoey. 1995. "Collective Restraint in a Social Dilemma Situation: The Influence of Procedural Justice and Community Identification on the Empowerment and Legitimation of Authority." *Journal of Personality and Social Psychology* 69: 482-497.

Tyler, T. R., P. Degoey, and H. J. Smith. 1995. *Understanding Why the Justice of Group Procedures Matters.* Unpublished manuscript. Berkeley: University of California.

———. 1996. *Journal of Personality and Social Psychology* 70(5).

Tyler, T. R., Y. J. Huo, and H. J. Smith. 1995. *Social Identity and Self-Esteem: Must Social Identity Information be Used Comparatively?* Unpublished manuscript. Berkeley: University of California.

Tyler, T. R., and E. A. Lind. 1990. "Intrinsic Versus Community-Based Justice Models: When Does Group Membership Matter?" *Journal of Social Issues* 46: 83-94.

———. 1992. "A Relational Model of Authority in Groups." Pp. 115-191 in *Advances in Experimental Social Psychology*, Vol. 25, edited by M. Zanna.

Tyler, T. R., E. A. Lind, and Y. J. Huo. 1995. *Culture and Reactions to Authority: Social Categorization Effects on the Psychology of Authority.* Unpublished manuscript. Berkeley: University of California.

Tyler, T. R., E. A. Lind, K. Ohbuchi, I. Sugawara, and Y. J. Huo. 1995. *Conflicts With Outsiders: Disputing Within and Across Cultural Boundaries.* Unpublished manuscript. Berkeley: University of California.

Tyler, T. R., and K. M. McGraw. 1986. "Ideology and the Interpretation of Personal Experience: Procedural Justice and Political Quiescence." *Journal of Social Issues* 42: 115-128.

Tyler, T. R., and G. Mitchell. 1994. "Legitimacy and the Empowerment of Discretionary Legal Authority: The United States Supreme Court and Abortion Rights." *Duke Law Journal* 43: 703-815.

Vermunt, R., A. Wit, K. van den Bos, and A. Lind. 1993. *The Effect of Inaccurate Procedure on Protest: The Mediating Role of Perceived Unfairness and Situational Self-Esteem.* Unpublished manuscript. Leiden University, The Netherlands.

Williamson, O. E. 1993. "Calculativeness, Trust, and Economic Organization." *Journal of Law and Economics* 36: 453-500.

Wit, A. P. 1989. *Group Efficiency and Fairness in Social Dilemmas: An Experimental Gaming Approach.* Doctoral dissertation. Groningen, The Netherlands: University of Groningen.

Yamagishi, T. 1986a. "The Provision of a Sanctioning System as a Public Good." *Journal of Personality and Social Psychology* 51: 110-116.

———. 1986b. "The Structural Goal/Expectation Theory of Cooperation in Social Dilemmas. Pp. 51-87 in *Advances in Group Processes*, Vol. 3, edited by E. Lawler. Greenwich, CT: JAI Press.

Yamagishi, T. 1988. "Seriousness of Social Dilemmas and the Provision of Sanctioning System." *Social Psychology Quarterly* 51: 32-42.

PHYSICAL ABILITY AS A DIFFUSE STATUS CHARACTERISTIC:
IMPLICATIONS FOR SMALL GROUP INTERACTION

Nancy L. Eiesland and Cathryn Johnson

ABSTRACT

In this chapter we argue that physical ability is a diffuse status characteristic in our society. We demonstrate this by examining negative stereotypes of and attitudes toward people with physical disabilities, the prevalence of interpersonal and institutional discrimination against the physically disabled, and preliminary evidence that demonstrates an association between physical ability and general expectation states. Since states of physical ability, namely able-bodied and disabled, are differentially valued in society, they also have implications for interaction in task groups. To address this, we discuss the role of categorical cues and task cues in the activation of physical ability in task groups. We provide some preliminary predictions of the effects of physical ability on the development of performance expectations. Finally, we suggest interaction techniques developed for other status characteristics may be adapted to reduce undesirable features of status generalization based on physical ability.

Advances in Group Processes, Volume 13, pages 67-90.
Copyright © 1996 by JAI Press Inc.
ISBN: 0-7623-0005-1

In recent years, people with disabilities have begun to move in greater numbers into the mainstream of economic, social, and educational life in the United States. This movement has been bolstered by the enactment of the Americans with Disabilities Act (ADA) which equates denying people with disabilities access to employment, buildings, public accommodations, transportation, or communication services with denying access to someone on the basis of race, religion, or gender. Though progress has been made in gaining access to opportunities for people with disabilities, stereotypes about people with disabilities often make interactions between the disabled and non-disabled tense and mutually unsatisfactory.[1] For example, research demonstrates that levels of stress in interaction are heightened for both people with physical disabilities and the able-bodied.[2] In addition, attribution of physical disability often leads to lower expectations of competence.[3]

In this chapter we show how physical ability is a relevant social category that attenuates social status for individuals who are labeled physically disabled. Specifically, we argue that physical ability is a diffuse status characteristic in our society and, therefore, has implications for interaction in small task groups. Status characteristics theory identifies diffuse status characteristics, such as race, gender, sexual orientation, and age, as having at least two states that are differentially evaluated in society in terms of value and esteem and are related to distinct sets of specific and general expectation states (Berger, Conner, and Fisek 1974; Berger, Fisek, Norman, and Zelditch 1977). We contend that physical disability is the low status category and able-bodied is the high status category for the diffuse status characteristic physical ability.[4] Viewing physical ability as a diffuse status characteristic suggests interaction techniques developed for other status characteristics may be adapted to overcome negative features of status generalization based on physical ability. We explore this possibility at the conclusion of the chapter.

Physical ability as a diffuse status characteristic is also theoretically interesting and complex for several reasons. First, although disability is a stigmatized social condition, all people with disabilities are not rejected equally (Goffman 1963; Hahn 1988; Meyerson 1988; Murphy 1995). Substantial differences exist in the degree of public comfort with people with disabilities. People with some disabilities, such as facial disfigurement and seizure disorders, are kept at a greater social distance than individuals from the least accepted ethnic minority group (Harasymiw 1971; see also Harasymiw, Horne, and Lewis 1976; Schneider and Conrad 1983). High school, college, and graduate students and health-care workers consistently rejected intimacy with persons of all types of disabilities, though persons with cerebral palsy, paraplegics, dwarfs, epileptics, and hunchbacks were deemed unsuitable even as neighbors (Tringo 1970). Facial disfigurement is the physical disability which arouses the greatest discomfort.[5] Seventy-two percent of the U.S. population say they are not very comfortable interacting with people who are facially disfigured. About two-thirds of the population are uncomfortable interacting with people who are deaf. Half of the U.S. population is uncomfortable interacting with people who are blind, and slightly less than half are ill at ease with people who are in wheelchairs (Louis Harris and Associates 1991). Thus, the type of disability is important in considering social acceptance and participation.

Second, individuals use different types of cues to identify physical ability, some of which are directly related to perceived competence on the task while other cues are only indirectly related. For example, some cues such as speech style (e.g., speech impairments), facial appearance (e.g., being blind), and demeanor (e.g., physical contortions) that may indicate physical ability will be directly relevant to the outcome of the task (either success or failure). The presence of these cues will result in much lower expectations of competence. Other cues, such as use of wheelchair, braces, and a single hearing aide, that may also indicate physical ability will have more indirect relevance and will result in only slightly lower expectations of competence. By way of illustration, people with paraplegia are often perceived as more competent than persons with other disabilities, such as muscular dystrophy or rheumatoid arthritis, because they appear "normal" except for their inability to walk.

Third, there is the possibility that attractiveness will be invoked simultaneously when some types of cues for physical ability are present. Substantial research shows that attractiveness pervades social relationships, resulting in unequal treatment for the unattractive and promoting an expectation of high levels of competency for the attractive (Dion, Berscheid, and Walster 1972; Jackson, Hunter, and Hodge 1995; Saxe 1979; Webster and Driskell 1983). Attractiveness in our society is associated with the acquisition of a higher level of intangible social goods, such as honor, attention, and esteem, as well as tangible social resources, such as employment (Schuler and Berger 1979; Umberson and Hughes 1987; Webster and Driskell 1983) In many cases, individuals with physical disabilities such as facial deformities will also be perceived as unattractive, suggesting that they are likely to be doubly stigmatized (Fine and Asch 1988; Unger, Hilderbrand, and Mardor 1982).[6] Evaluation of their competence may stem from a combination of cultural stereotypes about people with disabilities and the physically unattractive.

In this work, then, we argue that physical ability is a diffuse status characteristic in our society and, therefore, has implications for social interaction in informal task groups.[7] To present our argument, we divide this chapter into four sections. In the first section, we present a brief summary of status characteristics theory, with particular focus on the extension of this theory concerning the relationship between status cues and status in small groups (Berger, Webster, Ridgeway, and Rosenholtz 1986; Ridgeway, Berger, and Smith 1985). In the second section we discuss physical ability as a diffuse status characteristic, identifying two sorts of negative stereotypes that are associated with people with disabilities. These stereotypes are also related to discrimination against people with disabilities, both interpersonal and institutional. In section three we explore how physical ability as a diffuse status characteristic is activated in task groups through the use of categorical and task cues, paying special attention to paths of relevance to task performance. Also discussed is the concomitant activation of physical attractiveness in some circumstances. Finally, predictions for interaction in groups where members differentiate on physical ability are offered in section four. The work concludes with discussion of strategies for the amelioration of interaction disability.

STATUS CUES AND EXPECTATION STATES

The theory of status characteristics and expectation states addresses the creation of power and prestige orders in small task groups under specific scope conditions.[8] It seeks to determine the processes by which participants are given opportunities to contribute to group tasks and to gain status within the group (Berger and Conner 1974; Berger, Wagner, and Zelditch 1985). The power and prestige orders are measured by group members' action opportunities, performance outputs, reward actions, and influence. The theory explains the emergence of these orders by proposing that group members use status characteristics to evaluate each other's probable competence within the group. Research has demonstrated that status characteristics impact an individual's likely participation and, therefore, influence in the group (Meeker and Weitzel-O'Neill 1977; Ridgeway 1982).

The theory delineates two categories of status characteristics: specific and diffuse. Specific status characteristics refer to an individual's capacity to perform specific tasks pertinent to the group's assignment (e.g., reading or math ability). Diffuse status characteristics are individual variations on which particular social groupings qualitatively differentiate, such as gender (Meeker and Weitzel-O'Neill 1977; Ridgeway 1978), race (Cohen and Roper 1972), ethnic identities (Cohen and Sharan 1980), physical attractiveness (Landy and Sigall 1974; Webster and Driskell 1983), and sexual orientation (Johnson 1995). Diffuse status characteristics become associated with the distribution of intangible social goods such as honor, respect, and esteem, as well as with opportunities to contribute to the group task (Berger, Wagner, and Zelditch 1985, pp. 12-13).

A diffuse status characteristic is said to be activated, that is, become salient, when group members differentiate on the characteristic or when the task is related to the characteristic (e.g., gender-stereotyped tasks) and if nothing explicitly makes the characteristic irrelevant to the group task. In addition to these criteria of differentiation and relevance, the referential structure may be an important factor in determining when a diffuse status characteristic is activated (Ridgeway 1988). For example, if group members do not differentiate on a characteristic but collectively contrast with the authority structure, the characteristic may be activated. Once a status characteristic is activated, the cultural assumptions associated with the characteristic become salient. These views, in turn, set the foundation for the expectations members form for their own and other's performance at the task.

Expectations based on diffuse status characteristics become the basis for a self-fulfilling prophecy. On the basis of a diffuse status characteristic, a high status group member expects of herself/himself comparatively greater competence at a task. Likewise other group members expect greater competence from her/him and afford the individual more action opportunities and performance displays. Hence, the high status individual initiates more and, therefore, becomes more influential. Similarly, a low status member has low self-expectations resulting from the socially devalued diffuse status characteristic. Other group members also expect the person to be relatively less competent. Hence, the low status individual initiates less, is afforded fewer opportunities for interaction by other group members, and is, therefore, less influential within the group. Socially devalued diffuse status characteristics thus are translated into low status within the group through the evaluations of self and others which produce generalized performance expectations (Berger, Wagner, and Zelditch 1985).

Persons in the group who have low status due primarily to diffuse status characteristics experience "interaction disability" (Cohen 1972, 1993; Ridgeway 1982). Interaction disability is a fundamental inequality which precludes the low status individual from achieving influence within the group. Cohen (1972) identified "interracial interaction disability" as the tendency for interaction between blacks and whites to become strongly related to the race of the interactants. Cohen noted interracial interaction disability occurs when blacks and whites working together exhibit implicit expectations for superior performance and greater participation by whites in comparison to blacks. Hence, a recursive cycle of superior-inferior relationship of whites and blacks is enacted in the interaction process whereby whites expect inferior performance by blacks; the blacks internalize the whites' evaluations and, hence, fulfill those expectations of inferiority (Cohen 1972). Ridgeway (1982) extended interaction disability to include women, other minorities, and disadvantaged groups in society. She noted that interaction disability is a function of status generalization which happens when group members, in the absence of task-specific information, form their performance expectations on the basis of diffuse status characteristics, following the "burden of proof" approach. The burden of proof argument states that unless their inapplicability is demonstrated or justified, status characteristics will be applied (Berger, Wagner, and Zelditch 1985). The burden of proof argument has been specifically demonstrated for the following diffuse status characteristics: gender, race, age, military rank, and educational attainment (Berger, Wagner, and Zelditch 1985).

Ridgeway and colleagues (1985) and Berger and associates (1986) extended status characteristics theory by suggesting how different types of status cues are related to status in task groups. Status cues indicate the different social statuses and task ability individuals possess. They are observable signs of appearance, behavior, or surrounding possessions that group members use to infer each other's status in the group. Berger and colleagues (1986) distinguish between two types of status cues in their typology: categorical cues and task cues. Categorical cues refer to aspects about a person's appearance, behavior, or possessions that provide information on the social groups to which he or she belongs in the larger society. Group members use these cues to identify states of status characteristics possessed by each other; these cues activate beliefs and stereotypes associated with these characteristics (Berger et al. 1986, p. 13). Examples of categorical cues are accent, skin color, dress, diplomas on the wall, or statements made about social position outside the group such as, "I graduated from Princeton" (Berger et al. 1986, p. 6).

Task cues are nonverbal or verbal behaviors that indicate how well group members are performing and will continue to perform in the immediate group situation (Ridgeway et al. 1985, p. 963). They indicate the level of general problem-solving ability members possess that is relevant to the group task. Members use this information to infer how well each member is doing and will do at the task. Examples of task cues are eye gaze, body posture, response latency, tone of voice, as well as declarations that directly refer to a member's ability such as, "I just happen to know how to do this" (Berger et al. 1986, p. 6).

Categorical and task cues are further divided into two dimensions: indicative and expressive cues. Indicative cues are signs, objects, or statements that are directly presented which indicate possession of states of a status characteristic (categorical indicative cues) or states of abilities and competence (indicative task cues). Indicative categorical cues include diplomas, licenses, and statements such as, "I am a nurse." Indicative task cues are statements such as, "I know how to solve this problem," or "I don't know anything about this type of problem" (Berger et al. 1986, p. 4).

Expressive cues are nonverbal and verbal behaviors "given off" (Goffman 1959) during interaction that provide information about the diffuse status groups to which a member belongs (expressive categorical cues) or information about abilities or skills of the group member (expressive task cues). The former includes such things as accent, dress, word usage, and skin color. Examples of the latter are duration of eye contact, speech speed, tone of voice, speech fluency and hesitancy, posture, and maintaining a minority position (Berger et al. 1986, p. 6).

Categorical cues, both indicative and expressive, are most likely used by actors in the beginning of their interaction to determine salience of diffuse status characteristics. They likely play a major role in identifying these status characteristics and each member's relation to them (Ridgeway et al. 1985). Once activated, general performance expectations associated with the status characteristic are set into motion, affecting members' level of participation and influence within the group.

According to Ridgeway and colleagues (1985) and Berger and associates (1986), in peer groups where members do not initially differentiate on status characteristics, certain task cue behaviors (e.g., speaking fluently in a confident tone and using direct eye contact) are associated with attributions of competence and, therefore, lead to high expectations for those members displaying them; other task cue behaviors (e.g., speaking hesitantly with a submissive tone) lead to low expectations. These performance expectations then affect the development of the status hierarchy, where members with high expectations will participate more and will be more influential than members with low expectations. Once the hierarchy is formed, high status members continue to use those task cues associated with a high level of competence, while low status members continue to use task cues associated with a low level of competence, thereby maintaining the status hierarchy.

In a group of status unequals, however, where members initially differentiate on status characteristics, members' performance expectations for self and other form on the basis of these salient characteristics. These differences in performance expectations, then, are expressed in differential level of task cues, where those with high expectations display task cues associated with high levels of competence; those with low expectations display task cues associated with low levels of competence. There is some empirical evidence to support this argument about the relationship between task cues and status in groups of status unequals (Dovidio, Brown, Heltmann, Ellyson, and Keating 1988; Ridgeway et al. 1985).

Finally, Berger and colleagues (1986) argue that, if for some reason the differentiation in task cues is incongruent with the differentiation in categorical cues, then both sets of cues will be used to determine the members' expectations and behavior (Wagner and Berger 1993). For example, a black man in a racially mixed group who speaks fluently in a firm, confident tone possesses the low state of a categorical cue (black skin), yet possesses the high state of task cues (speech cues). This situation illustrates a status inconsistent situation. In this case, information from both types of cues will be combined in the formation of expectations.[9] However, task cues give information about abilities on the immediate group task; categorical cues indicate which states of diffuse status characteristics members possess that become relevant, usually indirectly, to the task. Consequently, the strength of relevance of task cues to perceived task performance is greater than that of categorical cues; that is, task cues have a shorter path of relevance to perceived task performance than categorical cues (Wagner and Berger 1993). Therefore, task cues are more likely to have a greater effect on performance expectations than categorical cues when they are inconsistent. Sometimes, however, task cues that signal high status may be misinterpreted by others in the group as "uppity" or "bossy" rather than as confident or assertive when displayed by members who possess low states of diffuse status characteristics.

PHYSICAL ABILITY AS A DIFFUSE STATUS CHARACTERISTIC

According to expectation states theory, an attribute can be a diffuse status characteristic if its states are differentially evaluated in society, so that one state is perceived as inferior to the other and the states are related to sets of specific and general expectation states (Berger, Rosenholtz, and Zelditch 1980). To demonstrate that states of physical ability, disabled, and able-bodied are differentially evaluated in our society in terms of esteem, opportunity, and value with disabled persons being less valued than able-bodied, we examine general stereotypes and attitudes toward people with disabilities and discuss interpersonal and institutional discrimination against the disabled.

In our society physical ability is categorized nominally; that is, the perception is that people differentiate on the characteristic categorically rather than in a graduated or ordinal way (Ridgeway 1991). In actuality, a person's state of physical ability is not a nominal category; rather it lies on a continuum which has been dichotomized into disabled and able-bodied (cf. Nagi 1979). This dichotomy has been constructed and upheld, in part, through governmental definitions of disability for purposes of federally-funded public assistance (Liachowitz 1988; Roth 1983) and medical definitions of organic variance (Albrecht 1992; Mashaw 1979; Safilios-Rothschild 1981). Furthermore, in contrast to most other status-valued categories, all able-bodied people have some probability of involuntarily becoming disabled during the course of their lifetimes. For example, though individuals cannot suddenly become black, change their ethnic origin or gender, or (ordinarily) dramatically improve their physical appearance, the majority of people with disabilities acquire their impairment after birth. Fewer than 15 percent of people with disabilities were born with their disability (Shapiro 1994). Thus for most people with disabilities, physical disability is an achieved status.

Differentiation on the basis of physical ability has also occurred because of identifiability, that is, the existence of characteristics that permit disabled individuals to be distinguished from the rest of the population (Goffman 1963; Fine and Asch 1988; Katz 1981; Kleck 1968). These characteristics may include physical or behavioral indicators and constitute status cues, that is, carriers of social information which activate status-organizing processes (Berger et al. 1986). Status cues of physical impairment, for example, use of a wheelchair, facial deformity or slurred speech, activate cultural beliefs about people with disabilities. Wright (1980, 1983) argues further that the presence of visible disability tend to "spread" to activate judgments by others of characteristics which have no necessary relationship to an individual's impairment. This "spread" can be either positive, that is, inferring such attributes as inner wisdom and courage, or negative, that is, indicating mental deficiency or rigidity (cf. Mussen and Barker 1944). In either case, cues of disability influence expectations others have of the person and his or her life opportunities, aspirations, and competence.

Further evidence indicates that beliefs about physical disability shape, not only the degree to which status cues of physical impairment are isolated from "normal" behavior, but also the actual perception of status cues. For example, negative attitudes toward persons with disabilities are often intensified because of overestimation of the individual's defect or the unattractiveness of her/his physical appearance (Ladieu, Adler, and Dembo 1948). Because physical disability is a devalued status in U.S. society and is clearly salient for the non-disabled and the disabled, people with disabilities are often stigmatized in interaction and subject to stereotypical representation.

Stereotypes About People With Physical Disabilities

Stereotypes regarding people with disabilities are of two dominant types: (1) stereotypes that depict people with disabilities as freaks or objects of fear and (2) stereotypes that depict people with disabilities as inspirational or objects of pity (Longmore 1985; Shapiro 1994; Zola 1985).[10] Concerning the first type, the preponderance of evidence suggests that persons with physical disabilities are generally regarded as socially deviant and feared by society (Gazsi 1994). Mass media has mirrored and shaped these attitudes. Film, television, and comics stereotype people with disabilities, most often portraying them as villains, monsters, or self-pitying neurotics (Longmore 1985). Often these stereotypes present people with disabilities as individuals who display self-loathing and a keen desire for revenge (Bogdan, Bilken, Shapiro, and Spekloman 1982). For example, the Joker, Penguin, and Two-Face of *Batman, Batman Returns*, and *Batman Forever* fame are "monsters" whose evil deeds can be traced to their bitterness at being disabled. Likewise the portrayal of a blind war hero in *Scent of a Woman* trades on cultural stereotypes of the disabled as tragic victims as well as bitter avengers. Repeated presentations of negative images of people with disabilities leads to negative expectations of interactions. Especially troubling in regard to this stereotype is the view that physical disability causes or promotes social pathology (see Gerschick and Miller 1994).

Not surprisingly, Berry and Jones (1991) found that the majority of non-disabled college students saw people with disabilities as threats to their beliefs in a just world. According to Goffman (1963), most stigmas, including physical disability, have an element of threat associated with them. These threats vary according to the stigma. The threats associated with the stigma of disability may include confounding a belief that innocent people do not suffer, engendering fear of sudden impairment or producing anxiety about being drawn into another person's dependency (Katz 1981, p. 3). Goffman theorizes that the effect of stigma becomes attached to other non-stigmatized people after deep interrelationship. It is because of perceived threat that others seek to avoid the stigmatized person.

Hahn (1988) has slightly altered and extended Goffman's notion of threat to suggest that status cues of physical impairment evoke anxiety in non-disabled persons which in turn lead to differential treatment, social distance, and stigmatization. This anxiety, Hahn maintains, is closely associated with social valuation of appearance and autonomy. The specific anxieties evoked by these social values are "aesthetic" and "existential," respectively. Aesthetic anxiety has as its referent culturally normative images of human physique or behavior. Aroused by the appearance of people with visible disabilities, aesthetic anxiety is exhibited in the common desire not to associate or interact with individuals perceived as strange or ugly. Existential anxiety is triggered by the threat of potential loss of functional capabilities embodied by the disabled. Hahn argues that the social stigma of a disability fundamentally derives from the fact that functional impairments may interfere with important life activities. Existential anxiety is often articulated in the there-for-the-grace-of-God-go-I response to people with disabilities. Upon interacting with a person with a disability the non-disabled individual may internalize the threat of disablement and project their existential fears onto the disabled person. Existential anxiety and a concomitant belief in the disabled individual's superior coping skills are also related to the second stereotype.

This stereotype depicts people with disabilities as inspirational or objects of pity. This stereotype is behind most television telethons. "Jerry's Kids" on the annual Muscular Dystrophy Telethon—as well as numerous other telecasts—are smiling, cheery individuals whose pluck makes the tragedy of disability even more pitiable (Fine and Asch 1988; Shapiro 1994). This stereotype also reinforces the idea that people with disabilities are somehow separate from ordinary human existence. In this instance, however, they are removed from the realm of normality by being elevated to sainthood and perpetual childlike innocence. The telethon stereotype portrays people with disabilities as needing sympathy and charity. The implication of this is demonstrated by several questionnaire studies that have found that most adults tend to assign highly positive traits to persons with disabilities (Kleck 1968; Mussen and Barker 1944). College students who rated orthopedically impaired individuals on 24 personality characteristics described as more conscientious, self-reliant, kind, persistent, intelligent, original, unselfish, and religious. They were rated unfavorably in two characteristics: overly sensitive and lacking in social adaptability (Mussen and Barker 1944).

Jordan (1963) argues that since overt expression of negative attitudes toward the disabled is not sanctioned by the majority, the majority group's general behavior pattern toward the disabled takes on the form of overconcern and flattery, similar to the sympathetic attitudes directed toward women and the aged. Thus, Safilios-Rothschild (1981) suggests that prejudice and discrimination directed toward the disabled might be more difficult to overcome than that which is directed toward racial and ethnic minorities since it is camouflaged in an air of protectiveness and admiration.

Interpersonal and Institutional Discrimination

Differential valuation of able-bodied and disabled states in U.S. society has resulted in widespread discrimination against people with disabilities. The findings of a Louis Harris poll are summarized as follows: "being disabled ... means having much less education and money and ... less of almost everything in life than most other Americans" (Louis Harris and Associates 1986, p. 24). Discrimination against people with disabilities is evidenced by most, if not all, of our institutions, for example, education, transportation, economic, recreation, religious, and family (cf. Albrecht 1992, chap. 5). Likewise survey data shows that people who reject various racial, religious, or ethnic groups also tend to reject disabled persons (Chesler 1965; English 1977).

Supported by stereotypes, most non-disabled people consider those with disabling conditions unable to work. High unemployment rates among people with disabilities can be attributed partially to employers' attitudes toward people with disabilities. One study demonstrated that prospective employers ranked physically disabled individuals lower than all minority groups, senior citizens, student militants, and prison parolees as prospective employees (Colbert, Kalish, and Chang 1973). More than two-thirds of people with disabilities of working age are not working (Louis Harris and Associates 1986). Persons with disabilities are unemployed because the physically disabled are thought to be handicapped in their productive capacities and are "seen to as social liabilities" in work settings (Best 1967, p. 3). Underemployment is also prevalent among people with disabilities. Many disabled persons are segregated in low-paying sheltered workshops, where they have little opportunity for advancement (Wertlieb 1985).

These staggering unemployment and underemployment rates do not stem entirely from the lack of capacity or desire to work. Studies show that people with disabilities perform as well or better than able-bodied coworkers, and more than 90 percent of people with disabilities want to work. Two-thirds of the U.S. public believe that people with disabilities are discriminated against in equal access to employment (Louis Harris and Associates 1991, p. 39). A substantial minority of people with disabilities who are employed or willing and able to work confront discrimination, unfavorable attitudes, and physical barriers in the workplace. Three in 10 report encountering job discrimination (Louis Harris and Associates 1986).

Unemployment and underemployment have resulted in concomitantly high levels of poverty among people with disabilities (Barnartt and Christiansen 1985). People with disabilities have more than twice as high a poverty rate as other U.S. citizens (McNeil 1993). Further, in the area often most essential to their well-being—health care—people with disabilities experience unsettling discrimination. Only half of persons with severe disabilities have private health insurance (McNeil 1993), and Medicare and Medicaid—federal programs for the poor and uninsured—insure only about one-third of disabled people (Shapiro 1994). Discrimination continues when people with disabilities obtain medical care. Medical professionals often disregard the preferences and opinions of people with disabilities (Albrecht 1992; Gething 1992).

Violence and abuse against people with disabilities is also common. People with disabilities are one and a half times more likely to experience at least one incident of physical abuse than are able-bodied individuals (Sobsey 1994; Sobsey and Varnhagen 1991). Rates of severe, multiple victimizations of children with disabilities are twice as high as for the general population (Sobsey 1994; Westcott 1991). In a study of Canadian women, Doucette (1986) found that physical abuse was nearly twice as common among women with disabilities as women without disabilities. Given higher rates of abuse of women generally in the United States, it is reasonable to speculate that this ratio is applicable in our society as well (Todd 1984).

Particularly disturbing about these findings is that most abuse is perpetrated against people with disabilities by family members, caregivers, or healthcare professionals (Fine and Asch 1988, p. 22). Often people with disabilities find it difficult to remove themselves from abusive relationships because of physical dependence and lack of independent financial support.

Discrimination and violence against people with disabilities also affects their opportunities to find and maintain intimate relations. Fine and Asch (1988) note that people with disabilities are seen as unsuitable as either heterosexual and same-sex partners. In a study of friendship between disabled and non-disabled women, researchers found that the non-disabled person's fear of the dependency of the disabled individual is a major barrier to intimacy (Fisher and Galler 1988).

The Americans with Disabilities Act (ADA), which was enacted into law in 1992, is aimed at amelioration of the considerable discrimination experienced by people with disabilities. The law requires employers to make "reasonable accommodations" for disabled employees; to make new public transportation accessible; to prohibit public places, such as restaurants, hotels, theaters, stores, and museums, from discriminating on the basis of disability; and to ban job discrimination based on disability. Attempts to protect people with disabilities from some types of institutional discrimination are beginning to address the most egregious of the differential treatment experienced by people with disabilities. However, they continue to face both the reality of and the constant threat of interpersonal and institutional discrimination because of their physical disability. The widespread discrimination and the prevalence of negative stereotypes of and attitudes toward people with disabilities indicate that physical disability is significantly less valued than non-disability in our society.

Physical Ability and General Competence Expectations

In addition to being differentially valued in society, states of the characteristic physical ability also provide cues which influence impressions others form of an individual. These impressions extend to include general expectations of competence that have no necessary relationship with physical ability. Wright (1980) contends that the presence of observed physical disability often results in a "negative spread," that is, the additional attribution of mental and psychological deficiency in addition to the physical disability. A spread of judgment occurs when people who are physically impaired are perceived to be mentally impaired (cf. Gething 1985). Heumann (1992) reports her own experience of this effect. "We were learning at a very early age that if you had a speech difficulty ... it was unlikely that you would successfully make it through the [school] system" (p. 193). She notes that teachers generally mistook slow speech as mental retardation and spent little time working directly with these students (see also Davis and Marshall 1987). Such autobiographical accounts of teacher bias and spread of judgment are confirmed by Schloss and Miller (1982) who found that terming a physically disabled student "special," that is, special education, consistently decreased teacher expectations for academic performance.

Relationship between physical ability and general expectations of competence has also been found among health professionals. De Loach and Greer (1981) found that medical doctors, rehabilitation nurses, and other health care workers consistently underestimated the capabilities of people with physical disabilities. They found the following behaviors among professionals that suggest low opinion of a client's competence: interpreting behaviors, such as expression of autonomy (i.e., living alone) and resistance to medical treatment (i.e., refusal to attend therapy sessions), that are seen as "normal" in non-disabled individuals as abnormal when exhibited by people with disabilities; assuming that people with disabilities need psychological counseling more than non-disabled individuals; and underestimating their ability to understand "complicated" medical procedures (1981, pp. 39-43).

Gething (1992) tested the relationship between the presence of a visible physical disability and judgments by health professionals, including occupational and physical therapists, speech pathologists, nurses, and medical students. Using videotapes of an interview between a 20-year-old person applying for a job in a bank and a personnel officer of the same gender, the study manipulated three variables, that is, visible presence of disability (use of wheelchair), manner (shy, neutral, or brash), and gender. After viewing the interview, subjects completed a semantic differential questionnaire which is used to test strong emotions rather than carefully constructed opinions.

Each of the three variables had a significant independent effect on reported impressions. Results supported the contention that visible disability strongly influences impression formation. Negative evaluations of respondents went far beyond the observable disability to influence judgments about social and psychological adjustment and competency. Respondents formed more favorable general impressions of individuals without a visible disability than for those with a disability. Least favorable impressions were reported for women with a visible disability and brash persons with visible disabilities. Among health professionals, physical ability clearly influences general expectations of social competence.

Obviously additional research is needed to further support and elaborate our claim that physical ability is a diffuse status characteristic. At this early stage, however, we have established that physical ability is a diffuse status characteristic, linked to general competence expectations.

ACTIVATION OF PHYSICAL ABILITY

Given that physical ability is a diffuse status characteristic, how does it become salient in a group context? We begin this analysis by examining how both categorical and task cues may be used to activate physical ability. We then discuss how physical attractiveness may also become salient simultaneously with physical ability in some situations.

There are several ways that physical ability may be activated. Categorical cues, both indicative and expressive, may be used to make attributions about a person's physical ability (Berger et al. 1986). Indicative categorical cues may include use of international access symbols on clothing or other personal property that reference an identity as someone with a disability. Other indicative cues are statements an individual may use to self-identify as a person with a disability, for example, "As a disabled person, I ..." Expressive categorical cues that most likely indicate disability include (1) use of assistive ambulatory, auditory, or perceptual devices such as wheelchairs, leg or arm braces, hearing aids, or directional canes; (2) speech style, for example, slurred, slow, or hesitant speech; (3) appearance, for example, facial disfigurement or bodily deformity; and (4) mannerisms, such as, physical contortions, involuntary movements, or shaking.

This description of possible cues that may activate physical disability is complicated by the fact that some of the above expressive categorical cues overlap with what Berger and colleagues (1986) label expressive task cues. For example, slurred or impaired speech, physical contortions, and involuntary movements may identify an individual as disabled but may also infer a low level of competence. Other expressive categorical cues such as presence of wheelchairs or hearing aids may also identify an individual as disabled but do not necessarily infer a low level of competence.

Given that task cues have shorter paths of relevance to expected task performance than categorical cues (Berger et al. 1986; Wagner and Berger 1993), we argue that categorical cues for physical ability that are also considered task cues, for example, speech style and demeanor, will have a direct path of relevance to task performance and will, therefore, result in lower expectations of competence than will categorical cues that identify physical disability but are not also task cues, for example, use of wheelchairs and hearing aids (cf. Higgins 1980). These cues will have only an indirect path of relevance to perceived task competence and, therefore, will result in only slightly lower expectations of competence. This suggests that expressive task cues, in addition to categorical cues, may activate particular diffuse status characteristics if specific task cues are used to identify a person's state of the diffuse status characteristic.

Also, in some situations, when physical disability is activated, physical attractiveness may be activated simultaneously. An individual with physical impairment or disfigurement, such as facial scarring, morbid obesity, or physical contortions, for example, may also be considered relatively unattractive, increasing this person's expectation disadvantage in the task situation. Certain cues that indicate physical disability, then, may also activate attractiveness and, therefore, have implications for interaction.

In order to flesh out the implications of physical abiiity in task groups, we offer specific predictions for situations that vary in how physical ability is activated. The types of cues that activate physical ability and their paths of relevance to task performance are key to how physical ability affects interaction and the development of status hierarchies.

Predictions in Same-Sex Dyads

To examine how physical ability may operate in task groups, we describe four situations which vary in how physical ability is activated and then offer predictions for the effect of this characteristic on the development of performance expectations. For the sake of simplicity, we begin this analysis by examining the effect of physical ability in same-sex dyads only, where members are equal on all other status characteristics. We also assume that physical ability will have similar effects in both all-male and all-female dyads, since there is no reason to believe that this characteristic will operate differently based on gender.

Situation One

In situation one, physical ability is activated through indicative categorical cues only. There is no visible evidence of disability—only a symbol or statement that identifies an individual as disabled. In this case, physical ability should become salient in both male and female dyads. Since physical ability carries with it general expectations for performance that are incorporated into small task groups whenever physical ability is activated in the situation, both the able-bodied and disabled member should have higher performance expectations for the able-bodied member; that is, the disabled member should have an expectation disadvantage compared to the able-bodied member. Consequently, the non-disabled person should be likely to contribute more to the task, have more opportunities to contribute, receive more positive evaluations of their contributions, and be more influential in group decisions than the disabled member.

Situation Two

In situation two, physical ability will become salient only through the presence of expressive categorical cues that are not also considered expressive task cues. In this case, physical disability is visible through the presence of cues such as an ambulatory assistive device or hearing aides. These cues activate physical ability, and, therefore similar to situation one, the disabled member should have an expectation disadvantage compared to the able-bodied person.

Situation Three

In the third situation, expressive categorical cues, that are not also task cues, simultaneously activate both physical ability and another diffuse status characteristic, physical attractiveness. These cues include physical disabilities, such as post-stroke syndrome, facial scarring, or morbid obesity, that significantly affect "normal" physical appearance. In both male and female dyads, both physical ability and attractiveness will be activated where the disabled member will possess the low state of both characteristics. In this situation, we predict that the disabled member will have a moderately greater expectation disadvantage than in situations one and two, since the disabled member is low on states of two diffuse status characteristics rather than one.

Situation Four

Finally, in the fourth situation, physical ability is activated through expressive categorical cues that could also be considered expressive task cues, such as speech style and demeanor. Given that task cues have shorter paths of relevance to task performance than categorical cues, we argue that the disabled member should have a much greater expectation disadvantage in this situation compared to the disabled member in situations one and two. Categorical cues that are also directly related to perceived competence will have a much greater effect on "interaction disability" than categorical cues found in the first two situations.

We also predict that the disabled member will have a slightly greater expectation disadvantage than the disabled member in situation three. This happens because, in the third situation, both characteristics are only indirectly related to task performance through general expectation states; in this fourth situation the disabled member not only has categorical cues that indicate disability, but some of these cues are also directly related to perceived competence since they are also task cues.

In summary, the predictions from these situations when examined together are as follows:

1. Disabled members will have a significantly greater expectation disadvantage in situations where the expressive categorical cues displayed are also expressive task cues than in situations where the categorical cues displayed, both expressive and indicative, are not also task cues.
2. Disabled members will have a moderately greater expectation disadvantage in situations where the categorical cues displayed also activate physical attractiveness than in situations where the categorical cues displayed do not activate attractiveness.
3. Disabled members will have a slightly greater expectation disadvantage in situations where the expressive categorical cues displayed are also task cues than in situations where categorical cues are not also task cues but do activate physical attractiveness.

The above predictions derived from status characteristics theory point to the complexity of how physical ability may be activated in interaction and the reality of interaction disability for disabled persons interacting with able-bodied persons. Stigmatized or low-status individuals come to expect from themselves what non-stigmatized or high-status individuals presuppose. Hence, the recursive cycle of inferior-superior relations form the foundation of face-to-face interaction.

REDUCING INTERACTION DISABILITY

Several status amelioration strategies have been tested and, therefore, have implication for devising strategies in reducing interaction disability between people with disabilities and the able-bodied. Ridgeway (1982, 1987) has suggested that people who enter a group with a low external status characteristic, such as females in mixed-sex groups or African Americans in interracial groups, can use group-oriented motivation, in addition to competent task contributions, to overcome their fundamental inequality. Group motivation's primary impact will be to reduce status generalization effects of diffuse status characteristics for low status individuals (Ridgeway 1982, p. 80). Webster and Driskell (1978) tested strategies for status equalization among white and black female students in which they determined that interaction disability experienced by minority group members could be decreased only if performance expectations of both high and low status members changed and if the diffuse status characteristic—race—was explicitly argued to be irrelevant to the successful completion of the group task. They also noted that legitimation by authority figures and superior specific status characteristics was necessary for overcoming low status generalization.

Cohen and Roper (1972), Cohen (1972, 1993), and Cohen, Lotan, and Catanzarite (1988) also demonstrated that in a controlled setting some status equalization can be attained. Cohen and Roper (1972) attempted to counteract interaction disability among four-person interracial groups of junior high school boys. The experiment consisted of a training task, taught by the black subjects, and a criterion task in which the black subjects exhibited greater competence than the white subjects. The competence of the black subjects was amplified by the high status role (teacher) they assumed in comparison to the lower status role (student) assumed by the whites. The results demonstrated that the effects of a diffuse status characteristic could be altered by increasing the competence of low status members on a general performance expectation, as well as modifying the expectations of the high status members. Cohen and Roper noted that simple demonstration of equality of competence between the black and white subjects was not enough; exhibiting the unmistakable superiority of the black subjects was necessary for status equalization.

Cohen and colleagues' (1988) study of elementary children in nine bilingual classrooms revealed that a classroom management system which inculcated cooperative norms, for example, asking questions, listening, helping others without actually performing the task, explaining to others, showing other people how things work and giving other people what they need, increased the interaction rates for all status groups, although status differentiation was still evident. Cohen notes that it is easier to remove learning inequalities than it is to eliminate status inequalities. Over time, however, status change was evidenced as children reported the classroom as a "friendlier place." The interaction disability of low status children was lessened, and the disparity in the status hierarchy was narrowed, although an equal-status situation was not created.

Cohen manipulated classroom expectations primarily through consciousness-raising of the persons with the highest status, for example, instructing teachers to delegate authority, and by creating situations whereby the advancement of the group depended upon members using both high and low status individuals as resources. She noted that, despite a modest leveling of status among the children, expectations based on diffuse status characteristics continued to combine with specific status characteristics to redouble the benefits for high and medium status individuals, while minimally advancing the status of the initially low status children.

Cohen and associates (1988) underscored the importance of referent actors, for example, teachers, in boosting the participation of the low-status persons. Citing studies which reveal that students who perceive the teacher as sharing their own characteristics may raise the expectation of the student, depending on the competence of the referent (Humphreys and Berger 1981), she comments that the role of authority is not simply as an authorizing agent but also as a role model. Cohen (1993) emphasized the importance of long-term hands-on educational programs required to overcome attitudinal barriers and to develop skills and understanding among these individuals.

Finally, research has demonstrated (Jones et al. 1984; Makas 1988) that interaction between disabled and able-bodied persons can be made more comfortable through the exchange of information between the individuals. Makas (1988) suggested that communication of expectations is vital for successful intimate relationships between disabled and able-bodied persons. Hastorf, Wildfogel, and Cassman (1979) and Belgrave and Mills (1981) have found that, in casual conversation, acknowledgment of the disability by the disabled individual can reduce the able-bodied person's discomfort. However, non-disabled and disabled interactants both prefer that this acknowledgment be initiated by the other individual (Sagatun 1985). No matter who initiates the acknowledgment of a visible disability, the exchange is most effective in reducing conversational stress among actors when it occurs early in the interaction (Belgrave and Mills 1981; Hastorf, Wildfogel, and Cassman 1979; and Sagatun 1985).

When initiated by the person with the physical disability, humor can also be a means for reducing interaction tension. Cartoonist John Calahan (1989) and numerous comedians with disabilities who were featured in the public television documentary, *Look Who's Laughing*, make their livings lampooning the social tensions which surround people with disabilities. Using the case of wheelchair users,

Cahill and Eggleston (1994) reported that laughter and good humor is often used to allay anxiety and decrease mutual embarrassment between people with disabilities and the non-disabled.

Reducing interaction disability of people with disabilities in small groups with participants of differentiated physical ability could be accomplished by display of group-oriented motivation by participants with disabilities, legitimation of low-status individuals by authority figures, representation of people with disabilities in high status roles, inculcation of cooperative norms for all participants, and exchange of information related to physical ability initiated by the person with disability. Group members with disabilities may also use humor to decrease their interaction disadvantage.

CONCLUSION

Incorporating physical ability into discussion of status characteristics theory strengthens the theory's ability to account for power and prestige orders in small groups. In this work, we have argued that physical ability is a diffuse status characteristic in our society. Evidence substantiates our claims about the existence of stereotypes of and discrimination against people with disabilities based on their physical ability. Preliminary evidence further indicates a link between the states of physical ability and general competence expectation.

We have shown how indicative categorical and expressive categorical cues activate physical ability in task groups. The complexity of the expressive categorical cues for physical ability was noted. Some expressive categorical cues for physical ability (e.g., slurred or impaired speech, physical contortions, and involuntary movements) are also expressive task cues with direct paths of relevance, while other cues (e.g., use of wheelchair or a hearing aide) have a less direct path of relevance. Another intriguing consequence of the activation of physical ability is the possibility of the simultaneous activation of the characteristic—attractiveness—in same-sex dyads. Preliminary predictions that address the different scenarios of the activation of physical ability in same-sex dyads are provided. We then examine techniques developed for other status characteristics to reduce interaction disability. Initial suggestions are made for their possible adapation to overcome disadvantage in interaction based on physical ability. Examining how physical ability affects small group interaction should augment current understanding about the link between expressive categorical and expressive task cues. It should also extend our understanding of status clusters, such as attractiveness and physical ability (Berger, Wagner, and Zelditch 1985).

With the passage and phased-in access requirements of the Americans with Disabilities Act, new avenues of social participation for people with disabilities have opened. Higher rates of employment of people with disabilities with able-bodied individuals are expected within the next decade. Thus a concerted effort to assess the status of these individuals in task and social groups should now be undertaken. Examining physical ability as a diffuse status characteristic and identifying possible strategies for decreasing the interaction disability within groups that differentiate on physical ability establishes a basis for the empirical research.

NOTES

1.   Hereafter, "people with disabilities" will refer to people with physical disabilities.

2.   Able-bodied subjects report greater emotional distress, exhibit higher physiological arousal, for example, increased sweating, fidgeting, and decreased eye contact, show less physical movement, display less variability in their verbal behavior, express opinions that are less representative of their previously reported beliefs, and terminate interactions sooner when interacting with a subject who appeared disabled than when interacting with a able-bodied person (Kleck 1968; Kleck, Ono, and Hastorf 1966; Richardson, Hastorf, Goodman, and Dornbusch 1961). Similarly, people with disabilities experience a higher level of discomfort, for example, less smiling behavior, self-reports of discomfort, decreased eye contact, and shorter interactions, when interacting with able-bodied subjects, as opposed to individuals who appeared disabled (Comer and Piliavin 1972).

3.   Classifications and operational definitions for physical disability are numerous and vary with the context, for example, eligibility for workmen's compensation, political organizing, or medical rehabilitation. Physical disability here designates a chronic or persistent physical impairment which, despite treatment, results in functional limitation or disfigurement. It should be further noted that this or any definition of physical disability cannot be reified or abstracted from the social context (Whyte and Ingstad 1995; cf. Albrecht 1992, pp. 18-23).

4.   Though we note that mental and emotional disability may also be relevant for expectation status in small task groups, attention to those disabilities is beyond the scope of this work.

5.   The highest levels of discomfort are reported for interaction with people with mental illness, senility, and mental retardation.

6.   The low state of the diffuse status characteristic attractiveness may conversely activate physical ability as a diffuse status characteristic. For example, research has shown that unattractive physical characteristics can also lead to the attribution that an individual has a disability such as epilepsy (Hansson and Duffield 1976).

7.   Research addressing physical ability in England and Canada are also addressed.

8.   The scope conditions are as follows: (1) study groups are engaged in goal-oriented tasks, the outcomes of which can be categorized successful or unsuccessful; (2) the nature of the task must require the interactants to consider one another's behavior; (3) participants must be emotionally committed to successfully completing the task: (5) the competence involved must not have been previously specifically associated or dissociated from the diffuse status characteristic; and (6) there must be no other cause for differentiating between the participants other than this diffuse status characteristic (Cohen 1972).

9.   For discussion of how actors combine all the information on multiple status characteristics that have become salient and relevant to the task, see Berger, Fisek, Norman, and Zeldtich (1977).

10.   Stereotypes about people with disabilities are not limited to the two identified here. Haller (1995) identifies two other common stereotypical representations found in news stories, that is, the medical model and the business model. The medical model depicts people with disabilities as passively dependent on health providers for maintenance and cure. The business model depicts people with disabilities as costly to society and businesses in particular. These stereotypical representations, while important, are not as widespread as those which perceive people with disabilities as freakish and objects of fear or inspirational and objects of pity.

REFERENCES

Albrecht, G. L. 1992. *The Disability Business*. Newbury Park, CA: Sage.

Barnartt, S. N., and J. B. Christiansen. 1985. "The Socioeconomic Status of Deaf Workers: A Minority Group Perspective." *The Social Science Journal* 22: 19-32.

Belgrave, F. Z., and J. Mills. 1981. "Effect Upon Desire for Social Interaction With a Physically Disabled Person of Mentioning the Disability in Different Contexts." *Journal of Applied Social Psychology* 11: 44-57.

Berger, J., and T. Conner. 1974. "Performance Expectations and Behavior in Small Groups: A Revised Formulation." Pp. 87-109 in *Expectation States Theory: A Theoretical Research Program*, edited by J. Berger, T. Conner, and M. H. Fisek. Cambridge, MA: Winthrop.

Berger, J., T. Conner, and M. H. Fisek. 1974. *Expectation States Theory: A Theoretical Research Program*. Cambridge, MA: Winthrop.

Berger, J., M. H. Fisek, R. Z. Norman, and M. Zelditch, Jr. 1977. *Status Characteristics in Social Interaction: An Expectation States Approach*. New York: Elsevier.

Berger, J., S. Rosenholtz, and M. Zelditch, Jr. 1980. "Status Organizing Processes." *Annual Review* 6: 479-508.

Berger, J., D. Wagner, and M. Zelditch, Jr. 1985. "Expectation States Theory: Review and Assessment." Pp. 1-58 in *Status, Rewards, and Influence*, edited by J. Berger and M. Zelditch, Jr. San Francisco: Jossey-Bass.

Berger, J., M. Webster, C. Ridgeway, and S. Rosenholtz. 1986. "Status Cues, Expectation, and Behavior." Pp. 1-22 in *Advances in Group Processes* (Vol. 3), edited by E. Lawler. Greenwich, CT: JAI Press.

Berry, J. O., and W. H. Jones. 1991. "Situations and Dispositional Components of Reactions Toward Persons With Disabilities." *Journal of Social Psychology* 131: 673-684.

Best, G. 1967. "The Minority Status of the Physically Disabled." *Cerebral Palsy Journal* 28: 3-8.

Bogdan, R., D. Bilken, A. Shapiro, and D. Spekloman. 1982. "The Disabled: Media's Monster." *Social Policy* 13: 32-35.

Cahill, S. E., and R. Eggleston. 1994. "Managing Emotions in Public: The Case of Wheelchair Users." *Social Psychology Quarterly* 57(4): 300-312.

Callahan, J. 1989. *Don't Worry, He Won't Get Far on Foot*. New York: Random House.

Chesler, M. 1965. "Ethnocentrism and Attitudes Toward Disabled Persons." *Journal of Personality and Social Psychology* 2: 877-882.

Cohen, E. 1972. "Interracial Interaction Disability." *Human Relations* 25: 9-24.

———. 1993. "From Theory to Practice: The Development of an Applied Research Program." Pp. 385-415 in *Theoretical Research Programs: Studies in the Growth of Theory*, edited by J. Berger and M. Zelditch, Jr. Stanford, CA: Stanford University Press.

Cohen, E., R. Lotan, and L. Catanzarite. 1988. "Can Expectations for Competence be Altered in the Classroom." Pp. 27-54 in *Status Generalization: New Theory and Research*, edited by M. A. Webster and M. Foschi. Stanford, CA: Stanford University Press.

Cohen, E., and S. Roper. 1972. "Modification of Interracial Interaction Disability: An Application of Status Characteristic Theory." *American Sociological Review* 37: 643-657.

Cohen, E., and S. Sharan. 1980. "Modifying Status Relations in Israeli Youth: An Application of Expectation States Theory." *Journal of Cross-Cultural Psychology* 11: 364-384.

Colbert, J. N., R. A. Kalish, and P. Chang. 1973. "Two Psychological Portals of Entry for Disadvantaged Groups." *Rehabilitation Literature* 32: 194.

Comer, R. J., and J. A. Piliavin. 1972. "The Effects of Physical Deviance Upon Face-to-Face Interaction: The Other Side." *Journal of Personality and Social Psychology* 23: 33-39.

Davis, M., and C. Marshall. 1987. "Female and Disabled: Challenged Women in Education." *National Women's Studies Association Perspectives* 5: 39-41.

De Loach, C., and B. G. Greer. 1981. *Adjustment to Severe Physical Disability*. New York: McGraw-Hill.

Dion, K., E. Berscheid, and E. Walster. 1972. "What is Beautiful is Good." *Journal of Personality and Social Psychology* 24: 285-290.

Doucette, J. 1986. *Violent Acts Against Disabled Women*. Toronto: DisAbled Women's Network Canada.

Dovidio, J. F., C. E. Brown, K. Heltmann, S. L. Ellyson, and C. F. Keating. 1988. "Power Displays Between Men and Women in Discussions of Gender-Linked Tasks: A Multichannel Study." *Journal of Personality and Social Psychology* 55: 580-587.

English, R. W. 1977. "Correlates of Stigma Toward Physically Disabled Persons." Pp. 207-244 in *Social and Psychological Aspects of Disability*, edited by J. Stubins. Baltimore: University Park Press.

Fine, M., and A. Asch. Eds. 1988. *Women With Disabilities: Essays in Psychology, Culture, and Politics*. Philadelphia: Temple University Press.

Fisher, B., and R. Galler. 1988. "Friendship and Fairness: How Disability Affects Friendship Between Women." Pp. 172-194 in *Women With Disabilities: Essays on Psychology, Culture, and Politics*, edited by M. Fine and A. Asch. Philadelphia: Temple University Press.

Gazsi, S. 1994. *A Parallel and Imperfect Universe: The Media and People With Disabilities*. New York: Columbia University Press.

Gething, L. 1985. "Perceptions of Disability of Persons With Cerebral Palsy, Their Close Relatives and Able-Bodied Persons." *Social Science and Medicine* 30(6): 562-565.

————. 1992. "Judgments by Health Professionals of Personal Characteristics of People With a Visible Physical Disability." *Social Science and Medicine* 34(7): 809-815.

Gerschick, T. J., and A. S. Miller 1994. "Gender Identities at the Crossroads of Masculinity and Physical Disability." *Masculinities* 2(1): 34-55.

Goffman, E. 1959. *The Presentation of Self in Everyday Life*. Garden City, NY: Doubleday/Anchor.

————. 1963. *Stigma: Notes on the Management of Spoiled Identity*. Englewood Cliffs, NJ: Prentice Hall.

Hahn, H. 1988. "Politics of Physical Difference." *Journal of Social Issues* 44: 39-47.

Haller, B. 1995. "Rethinking Models of Media Representation of Disability." *Disability Studies Quarterly* 15: 26-30.

Hansson, R. O., and B. J. Duffield. 1976. "Physical Attractiveness and the Attribution of Epilepsy." *Journal of Social Psychology* 99: 233-240.

Harasymiw, S. 1971. "Attitudes Toward the Disabled as a Function of Prejudice Toward Minority Groups, Familiarity With the Disability, Age, Sex, and Education." *Dissertation Abstracts International* 32: 2482.

Harasymiw, S., M. D. Horne, and S. C. Lewis. 1976. "A Longitudinal Study of Disability Group Acceptance." *Rehabilitation Literature* 44(3): 31-33.

Hastorf, A. H., J. Wildfogel, and T. Cassman. 1979. "Acknowledgment of Handicap as a Tactic in Social Interaction." *Journal of Personality and Social Psychology* 37: 1790-1797.

Heumann, J. 1992. "Growing Up: Creating a Movement Together." Pp. 191-200 in *Imprinting Our Image: An International Anthology by Women With Disabilities*, edited by D. Driedger and S. Gray. Winnipeg, Canada: Gynergy Press.

Higgins, P. C. 1980. "Societal Reaction and the Physically Disabled: Bringing the Impairment Back in." *Symbolic Interaction* 3(1): 139-156.

Humphreys, P., and J. Berger. 1981. "Theoretical Consequences of the Status Characteristics Formulation." *American Journal of Sociology* 86: 953-83.

Jackson, L. A., J. E. Hunter, and C. N. Hodge. 1995. "Physical Attractiveness and Intellectual Competence: A Meta-Analytic Review." *Social Psychology Quarterly* 58: 108-122.

Johnson, C. 1995. "Sexual Orientation as a Diffuse Status Characteristic: Implications for Small Group Interaction." Pp. 115-137 in *Advances in Group Processes*, Vol. 12, edited by E. Lawler. Greenwich, CT: JAI Press.

Jones, E. E., A. Farina, A. H. Hastorf, H. Markus, D. T. Miller, and R. A. Scott. 1984. *Social Stigma: The Psychology of Marked Relationships*. New York: Freeman.

Jordan, S. 1963. "The Disadvantaged Group: A Concept Applicable to the Handicapped." *Journal of Psychology* 55: 313-322.

Katz, I. 1981. *Stigma: A Social Psychological Approach.* Hillsdale, NJ: Lawrence Erlbaum Associates.

Kleck, R. 1968. "Physical Stigma and Nonverbal Cues Emitted in Face-to-Face Interaction." *Human Relations* 21: 19-28.

Kleck, R., H. Ono, and A. Hastorf. 1966. "The Effects of Physical Deviance Upon Face-to-Face Interaction." *Human Relations* 19: 425-436.

Ladieu, G., D. Adler, and T. Dembo. 1948. "Studies in Adjustment to Visible Injuries: Social Acceptance of the Injured." *Journal of Social Issues* 4: 55-61.

Landy, D., and H. Sigall. 1974. "Beauty as Talent: Task Evaluation as a Function of the Performer's Physical Attractiveness." *Journal of Personality and Social Psychology* 29: 299-304.

Liachowitz, C. H. 1988. *Disability as a Social Construct: Legislative Roots.* Philadelphia: University of Pennsylvania Press.

Longmore, P. 1985. "Screening Stereotypes: Images of Disabled People." *Social Policy* 16: 31-37.

Louis Harris and Associates. 1986. *The ICD Survey of Disabled Americans: Bringing Disabled Americans Into the Mainstream.* New York: International Center for the Disabled.

_____. 1991. *Public Attitudes Toward People With Disabilities.* Washington, DC: National Organization on Disability.

Makas, E. 1988. "Positive Attitudes Toward Disabled People." *Journal of Social Issues* 44: 49-61.

Mashaw, J. L. 1979. "The Definition of Disability From the Perspectives of Rehabilitation." Pp. 160-167 in *Disability Policies and Government Programs,* edited by E. D. Berkowitz. New York: Praeger.

McNeil, J. M. 1993. *Americans With Disabilities: 1991-1992. U.S. Bureau of the Census.* Current Population Reports, P70-33. Washington, DC: U.S. Government Printing Office.

Meeker, B. F., and P. A. Weitzel-O'Neill. 1977. "Sex Roles and Interpersonal Behavior in Task-Oriented Groups." *American Sociological Review* 42: 91-105.

Meyerson, L. 1988. "The Social Psychology of Physical Disability: 1948 and 1988." *Journal of Social Issues* 44: 173-188.

Murphy, R. 1995. "Encounters: The Body Silent in America." Pp. 140-158 in *Disability and Culture,* edited by B. Ingstad and S. R. Whyte. Los Angeles: University of California Press.

Mussen, P. H., and R. G. Barker. 1944. "Attitudes Toward Cripples." *Journal of Abnormal and Social Psychology* 39: 351-355.

Nagi, S. Z. 1979. "The Concept and Measurement of Disability in Disability Policies and Government Programs." Pp. 1-15 in *Disability Policies and Government Programs,* edited by E. D. Berkowitz. New York: Praeger.

Richardson, S. A., A. H. Hastorf, N. Goodman, and S.M. Dornbusch. 1961. "Cultural Uniformity in Reaction to Physical Disabilities." *American Sociological Review* 26: 241-247.

Ridgeway, C. 1978. "Conformity, Group-Oriented Motivation, and Status Attainment in Small Groups." *Social Psychology* 47: 175-188.

_____. 1982. "Status in Groups: The Importance of Motivation." *American Sociological Review* 47: 76-88.

_____. 1987. "Nonverbal Behavior, Dominance, and the Basis of Status in Task Groups." *American Sociological Review* 52: 683-94.

_____. 1988. "Gender Differences in Task Groups: A Status and Legitimacy Account." Pp. 188-206 in *Status Generalization: New Theory and Research,* edited by M. Webster, Jr. and M. Foschi. Stanford, CA: Stanford University Press.

_____. 1991. "The Social Construction of Status Value: Gender and Other Nominal Characteristics." *Social Forces* 70: 367- 86.

Ridgeway, C., J. Berger, and L. Smith. 1985. "Nonverbal Cues and Status: An Expecation States Approach." *American Journal of Sociology* 90: 955-978.

90   NANCY L. EIESLAND and CATHRYN JOHNSON

Roth, W. 1983. "Handicap as a Social Construct." *Society* 20: 56-61.
Safilios-Rothschild, C. 1981. "Disabled Persons Self- Definition and Their Implications for Rehabilitation." Pp. 39-56 in *Cross National Rehabilitation Policies: A Sociological Perspective*, edited by G. Albrecht. Beverly Hills, CA: Sage.
Sagatun, I. J. 1985. "The Effects of Acknowledging a Disability and Initiating Contact on Interaction Between Disabled and Non-Disabled Persons." *Social Science Journal* 22(4): 33-43.
Saxe, L. 1979. "The Ubiquity of Physical Appearance as a Determinant of Social Relationships. Pp. 9-13 in *Love and Attractiveness: An International Conference*, edited by M. Cook and G. Wilson. Oxford, UK: Pergamon.
Schloss, P., and S. R. Miller. 1982. "Effects of the Labels 'Institutionalized' Versus 'Regular School Student' on Teacher Expectations." *Exceptional Children* 48(4): 361-362.
Schneider, J., and P. Conrad. 1983. *Having Epilepsy: The Experience and Control of Illness*. Philadelphia: Temple University Press.
Schuler, H., and W. Berger. 1979. "The Impact of Physical Attractiveness on an Employment Decision." Pp. 33-36 in *Love and Attraction: An International Conference*, edited by M. Cook and G. Wilson. Oxford, UK: Pergamon.
Shapiro, J. 1994. *No Pity: People With Disabilities Forging a New Civil Rights Movement*. New York: Times Books.
Sobsey, D. 1994. *Violence and Abuse in the Lives of People With Disabilities: The End of Silent Acceptance?* Baltimore: Paul H. Brooks Publishing Co.
Sobsey, D., and C. Varnhagen. 1991. "Sexual Abuse, Assault, and Exploitation of Individuals With Disabilities." Pp. 201-216 in *Child Sexual Abuse: Critical Perspectives on Prevention, Intervention, and Treatment*, edited by C. Bagley and R. J. Thomlinson. Toronto: Center for Human Development and Research.
Todd, A. D. 1984. "Women and the Disabled in Contemporary Society." *Social Policy* 14(4): 44-46.
Tringo, J. L. 1970. "The Hierarchy of Preference Toward Disability Groups." *Journal of Special Education* 4: 295-306.
Umberson, D., and M. Hughes. 1987. "The Impact of Physical Attractiveness on Achievement and Psychological Well-Being." *Social Psychology Quarterly* 50: 227-36.
Unger, R., M. Hilderbrand, and T. Mardor. 1982. "Physical Attractiveness and Assumptions About Social Deviance: Some Sex-by-Sex Comparison." *Personality and Social Psychology Bulletin* 8: 293- 301.
Wagner, D. G., and J. Berger. 1993. "Status Characteristics Theory: The Growth of a Program." Pp. 23-63 in *Theoretical Research Programs: Studies in the Growth of Theory*, edited by J. Berger and M. Zelditch, Jr. Stanford, CA: Stanford University Press.
Webster, M., and J. Driskell. 1978. "Status Generalization: A Review and Some New Data." *American Sociological Review* 43: 220-236.
_____. 1983. "Beauty as Status." *American Journal of Sociology* 89: 140-165.
Wertlieb, E. C. 1985. "Minority Group Status of the Disabled." *Human Relations* 38: 1047-1063.
Westcott, H. 1991. "The Abuse of Disabled Children: A Review of the Literature." *Child Care, Health, and Development* 17: 243-258.
Whyte, S. R., and B. Ingstad. 1995. "Disability and Culture: An Overview." Pp. 3-32 in *Disability and Culture*, edited by B. Ingstad and S. R. Whyte. Los Angeles: University of California Press.
Wright, B. A. 1980. "Developing Constructive Views of Life With a Disability." *Rehabilitation Literature* 1(11-12): 274-279.
_____. 1983. *Physical Disability: A Psychosocial Approach*. New York: Harper and Row.
Zola, I. 1985. "Depictions of Disability: Metaphor, Message, and Medium in the Research and Political Agenda." *Social Science Journal* 22: 5-17.

IDENTITY AND FRIENDSHIP:
AFFECTIVE DYNAMICS AND NETWORK FORMATION

Dawn T. Robinson

ABSTRACT

This paper examines goodness and power in a hypothetical friendship structure premised on the assumption that individuals become friends in order to increase opportunities for enacting valued identities. Affect control theory simulations generated a simulated friendship network among actors whose identities systematically varied in goodness, power, and expressivity. Analysis of this simulated network suggests that actors' goodness was related to their presence in cohesion-based subgroups, while actors' power was more important in determining structural equivalence among the hypothetical actors. Relations were homophilous with respect to goodness but not power or expressivity. Powerful people seemed more willing to befriend relatively weak others than relatively strong others.

INTRODUCTION

Individual choices of interaction partners are among the most fundamental of social phenomena. From a sociologist's point of view, there is little not affected by patterns

Advances in Group Processes, Volume 13, pages 91-111.
ISBN: 0-7623-0005-1

of social interaction. Such choices lead to decisions about hiring, dating, friendship, and marriage (among many others) and are linked to processes such as organizational affiliation (McPherson, Popielarz, and Drobnic 1991), attitude formation (Erickson 1988), and information transfer (Carley 1986), to name just a few.

How do we choose our friends? Certainly, choices of interaction partners are always from among constrained alternatives. Physical proximity strongly predicts the formation and strength of friendship ties (Festinger, Schacter, and Back 1950; Newcomb 1961; Verbrugge 1983). In addition, institutional structures place restrictions on interaction partner choices. For example, while an undergraduate student and a university president walk the same campus, a fellow student is a far more likely candidate for interacting with the young scholar than the university president. However, beyond the constraints of institutional and physical structures, there remains some volitional aspect of friendship formation. In this paper I briefly discuss several social psychological assertions about the manner in which we choose our friends and then consider some implications of one of those claims—namely, that individuals seek interactions with those who provide opportunities for maintaining valued identities.

IDENTITY AND FRIENDSHIP

Social psychologists' understandings of the relationship between self and society begin with social relationships. We typically think of the social self as deriving in large part from reflected appraisals of valued others (Cooley 1902; Mead 1934; Rosenberg 1979), from social comparisons (Rosenberg 1979), and from observations of our own actions toward others (Bem 1972). Consequently, a great deal of work recognizes the importance of social relations in the development and maintenance of social identities (e.g., Burke and Reitzes 1981, 1991; Hogg 1992; Serpe 1991; Serpe and Stryker 1987; Stryker 1981; Tajfel 1981, 1982; Tesser, Pilkington, and McIntosh 1989). Much of this work concerns identity as *object* rather than *agent*. Burke and Reitzes (1991, p. 239) point out, however, that to do a better job of understanding the effects of individual level processes on societal level outcomes, social psychologists must give more attention to the active self.

Social actors are often in a position to do behavioral work to negotiate identities. Among the behavioral strategies proposed is selective social interaction (e.g., Backman and Secord 1962; Robinson and Smith-Lovin 1992; Swann 1985, 1987). Through selective social interaction, actors can negotiate identities by choosing to interact with others who are likely to help produce specific outcomes.

Selective Interaction

To examine the implications of strategic action in choice of interaction partners, we must first consider possible motives for friendship formation. Empirical and theoretical work in social psychology suggest at least three alternative motives:

1. *People choose friends in order to gain access to valuable resources (social exchange).* Through both experimental and naturalistic studies, researchers have widely documented the involvement of network position in determining resources. Networks can serve as resources by providing access to information (e.g., Carley 1986), to social support (e.g., Suitor 1987; Suitor and Keeton in press), to influential others (see review in Campbell, Marsden and Hurlbert 1986) to trading power (e.g., Markovsky, Willer, and Patton 1988; Skvoretz and Willer 1991), and so on. Friends can help us achieve specific, identity-relevant goals either because of direct properties of themselves, or because of the their relations with others. Thus, one selective interaction strategy might be to choose friends on the basis of access to valued resources.

2. *People choose friends in order to enhance self-outcomes (self-enhancement).* Most of us are more attracted to those who like us than to those who do not (Berscheid and Walster 1978; Hays 1984). Self-enhancement theory (Taylor and Brown 1988) and self-evaluation maintenance theory (Tesser et al. 1989) suggest that this is part of a self-management strategy whereby we pursue more positive self views. Through selective interaction, we can manipulate our own reflected appraisals and produce selected referents for social comparison, weighting self-relevant evidence in our own favor.

3. *People choose friends in order to maintain valued identities (identity maintenance).* Symbolic interactionists (Cooley 1902; Heise 1979; Mead 1934; Stryker 1981) suggest that people strive for stable, coherent information about the self. According to this perspective, friendship choice can serve as an active strategy (in the same way described above) for eliciting consistent self-information.

Each of these identity-based motives is consistent with central theoretical traditions in social psychology. However, the first two motives for friendship choice have been more thoroughly explored in the social psychological literature. The present paper will focus on the third motive—identity maintenance—and explore some concrete implications of such a motive.

Affect control theory (Heise 1979; MacKinnon 1994; Smith-Lovin and Heise 1988) and identity theory (Burke 1980; Burke and Reitzes 1981, 1991; Stryker 1981) present symbolic interactionist formulations of identity in which the self is both object and agent. Both theories advocate measurement of identities based on semantic differentials. Both theories present cybernetic models in which there are forces promoting stability (Burke and Reitzes 1991). And, both theories imply that selection of friendships would be involved in identity formation and maintenance.

Identity theory describes identity-based action as deriving from commitment, which in turn arises through extensiveness and strength of connection to others who support that identity (Serpe 1991; Serpe and Stryker 1987). According to this theory, the frequency of role performance increases with the number of friends who value that role and the extent to which an individual values the friends that support that role. Empirical research in this paradigm typically has respondents identify their network alters as being associated with one or more specific identities. In contrast, the present paper considers the properties of others that make them likely to support various identities. Affect control theory is well-suited for this endeavor by virtue of being formally stated in a set of equations that allow computer simulation of social events to explore the implications of the theories' assumptions and empirical grounding.

FRIENDSHIP FORMATION AS AFFECT CONTROL

Affect control theory (Heise 1979; Smith-Lovin and Heise 1988) is a model of social interaction that provides a testable account of the way we interpret and create events to maintain fundamental meanings. Affect control theory focuses on the individual's definition of the situation, including the perception of self and others as occupants of various identities. As implied by its name, affect control theory posits a cybernetic model, or control system (Powers 1973), assuming that people operate in identities that carry *fundamental sentiments* that they attempt to control. Identities are typically viewed as schema-like structures, hierarchically organized within the self, and activated through particular settings and vis-à-vis particular interaction partners. Affect control theory emphasizes the affective component of these cognitive structures. Affective meanings associated with particular identities (fundamental sentiments) are represented by the dimensions of evaluation, potency, and activity (EPA) identified by Osgood, Suci, and Tannenbaum (1959). Other aspects of social interactions—social settings, behaviors, attributions, and emotions—are also identified in terms of their associations on these three dimensions.

Events (in the form of [actor] [verbs] [object]) create *transient impressions* due to the nature of the setting, the identities involved, and the behaviors performed. Transient impressions are the affective meanings associated with the particular combination of identities, settings, behaviors, and emotions in the current situation. Like fundamental sentiments, transient impressions are represented in affect control theory by their associations of evaluation, potency, and activity (EPA). For example, the identity *grandfather*, has an EPA profile[1] of (2.2, 1.3, -1.8), suggesting that within this culture we consider grandfathers to be very good, somewhat powerful but rather passive actors. When an actor's (situationally produced) transient impression differs from his or her (stable, culturally defined) fundamental sentiment, a *deflection* results. Actors then are motivated to construct new events that realign transient impressions with fundamental sentiments. Deflection is operationalized as the sum of the squared differences between the fundamental sentiment and the transient impression on each of the three dimensions (*e, p,* and *a*) over each of element in the event (actor, behavior, object, and setting). Thus, in the three *dimensional* semantic space, a deflection roughly corresponds to the distance between the fundamental sentiment and the transient impression.

Affect control theory presumes that this underlying control process is the same for everybody and that there is a large body of shared cultural meaning attached to common identities, situations, behaviors, and emotions. Evidence does suggest that within the United States these meanings are similar across socioeconomic status, age, and region (see review in Smith-Lovin 1990, p. 280). Differences in definition of the situation are modeled by differences in the labeling of behavior, identities, traits, and so on. For example, when a *father* asks his teenage *son* about his day at school, the father may view the situation as *father talks to son*, while the son may perceive the situation as *father interrogates son*. In this way, while the father and son have a similar understanding of the acts *talk to* and *interrogate*, they have very different understandings of the after school conversation.

Affect control theory is stated as a set of equations concerning the affective dimensions of elements of the social situation. These equations consist of impression-formation equations developed by Smith-Lovin (1987) and mathematical transformations of these equations to describe likely behaviors, emotions, and labeling (Heise 1991, 1992). The equations predicting impression-formation empirically derive from research using a sample of U. S. undergraduates.[2] The *impression formation* equations explicitly express the ways in which social events prompt affective reactions. The equations provide the detailed statement of affective dynamics that is necessary in order to examine implications of the affect control principle. To implement this principle, the impression formation equations are mathematically transformed into *impression management* equations. Operationalizing the affect control principle through the minimization of deflection, the impression management equations mathematically derive from the impression formation equations by solving for the event constructions that produce the lowest possible deflection. Using the EPA ratings of social interaction elements (e.g., identities and behaviors) as inputs and outputs, these impression management equations formally express the theory, including the affect control principle and the empirically derived statements about affective responses to events.

In summary, affect control theory uses information about the setting, actors, and behaviors in a social situation to predict information about impression formation and impression change. Actors resist change in their fundamental meanings. Such changes produce deflections—affective disruptions from their initial definition of the situation—that are signalled by the experience of emotion. Emotional experience communicates to the actor that deflection has occurred and prompts him or her to construct an event that would reduce that deflection and move impressions back toward his or her fundamental sentiments. The inputs and results of these equations are three-number profiles representing the ratings of evaluation, potency, and activity for each of the relevant social elements.

Affect Control and Interaction Choices

What does affect control theory have to say about actors' preferences for inter-action partners? Affect control theory predicts actors' choices of behaviors *during* interaction but does not predict *whether* two individuals will interact. However, predictions about interaction partner choices should be within the theoretical scope of affect control theory. For example, one answer is that nice people will interact with nice people. If we take the equation predicting the evaluation of an actor after a particular event (Smith-Lovin, 1987, p. 48) and drop all terms whose coefficients are less than 0.10, the equation reduces to:

$$A_e' = -0.10 + 0.47A_e + 0.42B_e - 0.11B_a + 0.13B_eO_e \qquad (1)$$

where A, B, and O refer, respectively, to properties of the actor, behavior, and object, and e, p, and a refer respectively to the evaluation, potency, and activity of the elements. The B_eO_e interaction term reveals that in addition to the nature of our behavior, the object (or recipient) of our behavior is important in determining how nice we are. Doing something nice to someone nice increases an actor's evalua-tion. In contrast, doing something nice to a negatively evaluated object mitigates, somewhat, the positive effect of the nice behavior.[3] So, actors striving to maintain positive identities should behave (nicely) toward nice people. Who are nice peo-ple? Again according to equation (1), nice people are those who do nice things (to nice others).

However, affect control theory predicts that only actors with nice identities will strive to maintain those positive associations. The more proximate goal is that of meaning maintenance. Within a particular social event, the deflection is given by,

$$D_i = (f_i - t_i)^2 \qquad (2)$$

where f represents the fundamental sentiment, t represents the transient impres-sion, and i represents the association of e, p, and a for each actor, behavior, object, and setting in the social event. According to Heise (1992), the "unlikelihood" (U) of an event is a function of the deflection it produces,

$$U = k + \sum w_i D_i \qquad (3)$$

where k is a constant and w is a vector of arbitrary weights. MacKinnon and Heise (1987) report empirical support for the proposition that people perceive events with high deflections to be more unlikely than events producing lower deflections.

This suggests a strategy for eliciting affect control predictions about the rela-tionship between identities and friendship choice. Relying on the affect control premise that individuals act to minimize deflection, we would surmise that when given a constrained choice between equally possible interaction partners, actors will choose the interaction predicted to produce the lowest deflection. Using this logic, Robinson generated affect control predictions about interaction partner choices simulating all possible interactions and searching for interactions producing the lowest deflection (Robinson 1992; Robinson and Smith-Lovin 1992). In other words, given an array of potential network alters who are equally possible interaction partners (with regard to physical and institutional constraints outside the affect control model), affect control theory predicts that actors will choose alters with whom interactions will produce acceptably low deflections.[4]

GOODNESS, POWER, AND EXPRESSIVITY IN A SIMULATED FRIENDSHIP NETWORK

As described earlier, affect control theory is based on the symbolic interactionist idea that individuals try to experience social life in ways that are consistent with their interpretations of their environments. Rather than grappling with the problem of measuring the qualitative richness of social cognitions, affect control theory uses affective measures to characterize these interpretations. Affect control theory assumes that all social cognitions evoke affective associations and that these associations can be indexed to a large degree on universal dimensions of response (MacKinnon and Heise 1993). To achieve this indexing, the theory relies on the semantic differential and exploits the work of Osgood and his colleagues (Osgood, May, and Miron 1975; Osgood, Suci, and Tannenbaum 1957) who have identified the dimensions of evaluation, potency, and activity as universal dimensions of meaning captured by the semantic differential.[5] These dimensions have been proposed to correspond with the social dimensions of status, (Kemper 1978; Kemper and Collins 1990), power (Kemper 1978; Kemper and Collins 1990), and Parson's concept of social expressivity (Heise 1987). Kemper and Collins (1990) argued that status and power are the fundamental dimensions of social life and deserve their places as such in our theories of both macro- and microstructures. Kemper and Collins relate Osgood and colleagues' potency dimension to control or dominance, which they call power, and they relate the evaluation dimension to acceptance and positive association. Traditional conceptions of status (see Ridgeway and Walker 1995), incorporating notions of esteem, deference, and influence, probably imply some combination of evaluation and potency.

How are these affective associations related to relationship choices? Since the 1930s and 1940s, sociometric techniques have been used to study power and popularity. Social network researchers have examined coerced and naturally occurring exchange, communication, and friendship relations to gain insight into the mechanisms of power, status, and gregariousness. In the following section I use simulated friendships to explore the dynamics of status, power, and expressivity in relationships among hypothetical actors whose goals are to maintain affectively stable identities.

Simulated Network Data

In order to make computation and interpretation using affect control equations more accessible to the researcher, Heise (1991; Heise and Lewis 1988) built these equations into a software program called INTERACT that allows researchers to simulate social situations and generate theoretical predictions about what behaviors are likely to occur, what emotions are likely to be experienced, and how actors and objects are likely to change their impressions of one another as a result of these behaviors and emotions. The predictions are reported in both numeric form—a three-number profile corresponding to the EPA rating of the concept (identity, behavior, etc.) and in common language form—a list of words (behaviors, identities, etc.) whose EPA ratings fall close to the predicted profile. These predicted behaviors, emotions, and identities come from a dictionary of words rated on EPA dimensions by a sample of U. S. undergraduates.

I searched this dictionary (Heise and Lewis 1988) for identities that were spread as evenly as possible across the semantic (EPA) space. To do this, I arbitrarily trichotomized each dimension into high (1 to 2), medium (-1 to 1), low (-2 to -1), and then chose 27 actors who had each possible combination of high-, medium-, and low-evaluation, -potency, and -activity. Table 1 shows the identity names that were used and their EPA profiles. Some areas in the semantic space are surprisingly bare of viable identities (Smith-Lovin 1990). These sparse areas frequently are filled in with mythical or fictitious identities (e.g., vampire, zombie). Accordingly, not only is the group of actors in this network unrepresentative of any naturally occurring group or population, some actors may not even be representative of any living *individual*.

Using INTERACT, I simulated the event *[Actor] Befriends [Object]* for all possible combinations of the 27 actors and objects (including friendships between actors with the same identity),[6] and recorded the deflections into a 27×27 asymmetric matrix. Because the (un-)likelihood of an event is presumed to be a function of the deflection (see equation [3]), high deflections suggest lower likelihoods of occurrence. To render the values more intuitively interpretable, I reversed the values in the matrix by subtracting each deflection from a constant ($k = 15$) to produce a "likelihood" matrix (shown in Table 2), with rows (i) and columns (j) where the elements refer to the likelihood of the event "i befriends j." For purposes of exploring affect control theory's predictions concerning identity-affect and the development of friendship networks, this likelihood matrix can be treated like a network matrix of tie "counts."

Goodness, Power, Expressivity, and Homophily

One of the most frequently invoked concepts in the research on friendship is homophily, the tendency of people to be friends with similar others (Feld 1981, 1982; Hallinan 1974; Johnsen 1986; Lazersfeld and Merton 1954; Marsden 1988;

Table 1. Social Identities in Simulated Network

Position in Semantic Space[1]		EPA Profile		
EPA Profile	Identity Name	Evaluation (nice-mean)	Potency (strong-weak)	Activity (lively-passive)
1. HHH	winner	1.53	2.17	1.63
2. HHM	surgeon	1.81	2.09	−0.21
3. HHL	grandfather	2.17	1.31	−1.77
4. HMH	playmate	1.92	0.46	1.95
5. HMM	darling	1.85	0.60	0.67
6. HML	grandmother	2.24	0.14	−1.92
7. HLH	baby	1.99	−2.56	2.49
8. HLM	maiden	1.63	−1.14	1.06
9. HLL	oldtimer	1.28	−0.82	−1.92
10. MHH	linebacker	0.14	2.06	1.92
11. MHM	sheriff	0.43	1.95	0.07
12. MHL	judge	0.89	2.34	−1.70
13. MMH	auctioneer	0.57	0.35	1.92
14. MMM	miner	0.07	0.28	0.28
15. MML	librarian	0.39	0.18	−1.53
16. MLH	newsboy	0.60	−1.31	1.63
17. MLM	midget	0.14	1.24	0.35
18. MLL	invalid	−0.57	−2.31	−1.35
19. LHH	brute	−1.95	1.70	1.92
20. LHM	mafioso	−1.77	2.09	0.32
21. LHL	vampire	−2.17	1.56	−0.57
22. LMH	maniac	−1.99	0.39	1.81
23. LMM	henchman	−1.85	0.50	0.28
24. LML	scrooge	−1.99	0.32	−1.99
25. LLH	crybaby	−1.67	−1.74	1.67
26. LLM	slave	−1.74	−2.38	0.11
27. LLL	zombie	−1.81	−1.95	−1.81

Note: Profiles taken from Heise and Lewis (1988).
Target criteria (approximately):
High = 1.5 to 2.5
Medium = -0.5 to 1.0
Low = -1.5 to -2.5

McPherson and Smith-Lovin 1986, 1987; Suitor and Keeton in press; Suitor, Pillemer, and Keaton 1995; Tuma and Hallinan 1979; Verbrugge 1977, 1979). Homogeneous relations can be either produced by or epiphenomenal to actors' motivation for such relations. Evidence suggests that both are true; homogeneous relations are in part a by-product of constrained opportunities (Feld 1981, 1982,

Table 2a. Simulated Friendship Network—Likelihood of [Actor] Befriends [Object]

Evaluation	High									Medium									Low								
Potency	High			High Medium			Low			High			Medium Medium			Low			High			Low Medium			Low		
Activity	H	M	L	H	M	L	H	M	L	H	M	L	H	M	L	H	M	L	H	M	L	H	M	L	H	M	L
	1	2	3	4	5	6	7	8	9	10	11	12	13	14	15	16	17	18	19	20	21	22	23	24	25	26	27
1	11.0	11.6	12.4	12.9	13.2	13.4	13.3	14.1	13.8	11.1	11.9	11.4	13.1	13.3	13.4	13.8	13.7	12.3	9.9	10.1	10.1	10.9	11.4	10.8	11.4	10.9	10.6
2	11.8	12.0	12.5	13.8	13.8	13.4	14.3	14.3	13.7	11.8	12.2	11.0	11.3	13.6	13.3	14.4	14.0	12.0	10.3	10.2	9.9	11.3	11.3	10.2	11.7	10.8	10.0
3	12.1	11.9	12.1	13.8	13.6	13.0	14.1	14.3	13.3	12.0	12.0	11.0	13.6	13.1	12.7	13.9	13.3	11.3	10.2	9.8	9.3	10.7	10.6	9.3	10.8	9.8	9.0
4	11.9	12.0	12.5	12.8	13.1	13.2	12.3	13.4	13.6	12.0	12.3	11.9	12.9	13.0	13.1	12.9	12.9	11.7	10.4	10.6	10.2	10.5	10.9	10.5	10.1	9.7	9.9
5	12.5	12.5	13.0	13.7	13.8	13.8	13.5	14.2	14.2	12.6	12.8	12.2	13.7	13.6	13.7	13.8	13.7	12.4	11.0	11.1	10.7	11.3	11.6	10.9	11.1	10.6	10.6
6	12.3	12.1	12.4	13.3	13.3	13.2	12.8	13.6	13.4	12.2	12.3	11.6	13.1	12.9	12.9	13.0	12.8	11.2	10.2	10.2	9.6	10.2	10.4	9.6	9.7	9.0	9.1
7	9.7	9.8	9.3	8.0	8.9	9.9	2.3	6.7	8.3	9.8	10.2	9.9	8.0	8.5	9.1	5.6	6.4	4.6	7.9	9.0	8.1	5.9	7.1	6.9	2.0	1.7	3.5
8	11.9	9.8	12.0	11.8	12.2	12.4	9.5	11.6	12.5	12.1	12.3	11.9	11.9	13.2	12.5	11.0	11.4	10.3	10.6	11.0	10.4	9.8	10.5	10.2	8.2	8.0	8.9
9	11.7	11.7	12.0	12.1	12.5	12.7	10.9	12.4	13.1	10.0	10.9	10.3	12.1	12.5	12.7	11.9	12.2	11.4	10.5	10.7	10.3	10.3	10.8	10.5	9.4	9.3	10.0
10	9.6	10.2	11.1	11.6	12.0	12.2	12.3	13.0	13.1	10.0	10.9	10.3	12.1	12.7	12.8	13.0	13.3	12.5	9.6	9.8	10.2	10.9	11.4	11.2	11.7	11.6	11.4
11	10.8	11.0	11.6	12.8	12.9	12.7	13.7	14.1	13.5	11.2	11.6	10.7	13.3	13.4	13.2	14.1	14.0	12.8	10.6	10.4	10.6	11.8	12.0	11.2	12.6	12.1	11.5
12	10.6	10.4	10.7	12.8	12.7	11.8	13.7	13.8	12.3	10.8	12.0	9.6	13.1	12.8	12.1	13.9	13.3	11.4	9.9	9.4	9.3	11.1	11.0	9.5	11.9	10.9	9.7

100

13	11.0	11.1	11.5	12.0	12.3	12.4	11.9	12.8	13.1	11.5	11.7	11.6	11.7	11.9	12.8	12.9	12.3	10.7	10.8	11.1	11.4	11.1	11.1	10.9	11.1
14	11.3	11.2	11.5	12.4	12.5	12.4	12.5	13.2	13.3	11.8	11.9	11.1	12.9	13.1	13.0	13.3	13.4	12.8	11.1	11.0	11.9	11.5	11.8	11.6	11.7
15	11.4	11.2	11.5	12.6	12.6	12.4	12.7	13.3	13.2	11.9	11.8	10.9	13.0	13.0	13.3	13.3	12.9	13.2	11.1	10.8	11.6	11.5	11.6	11.3	11.3
16	10.9	10.8	10.7	10.6	11.1	11.0	8.3	10.5	11.4	11.3	11.4	10.8	11.4	11.0	11.6	10.2	10.7	9.9	10.4	10.7	10.2	10.3	10.0	8.2	8.1	8.9
17	11.0	10.9	10.9	11.0	11.4	11.4	9.0	11.1	12.0	11.5	11.7	11.0	11.5	11.9	12.0	10.8	11.3	10.8	10.7	10.4	10.9	10.6	10.9	9.1	9.1	9.9
18	9.2	9.5	9.4	8.3	9.2	9.5	4.5	8.0	9.8	9.8	9.8	8.9	8.6	8.2	7.5	10.3	9.1	9.9	9.5	8.2	9.2	9.2	5.9	6.2	7.7
19	6.5	6.9	7.7	8.2	8.6	8.8	8.5	9.6	9.9	7.4	7.3	9.2	9.8	9.9	9.8	10.2	7.6	7.8	8.4	8.8	9.3	9.3	9.1	8.9	9.1
20	7.2	7.5	8.2	9.3	9.5	9.3	9.7	10.6	10.3	8.1	7.6	10.1	10.6	10.4	10.9	11.1	8.3	8.3	8.8	9.6	9.9	9.6	10.1	9.7	9.5
21	7.1	7.0	7.4	8.8	8.9	8.4	9.1	9.9	9.5	8.0	7.0	9.7	10.0	9.6	10.2	10.3	9.4	8.2	8.0	9.2	9.4	8.8	8.9	8.7
22	7.4	7.3	7.6	8.3	8.5	8.4	7.9	9.1	9.5	8.3	7.5	9.2	9.6	9.6	9.4	9.7	9.2	8.4	8.3	8.2	8.9	8.9	9.1	8.6	8.4	8.7
23	8.3	8.1	8.4	9.5	9.6	9.3	9.3	10.3	10.4	9.2	8.2	10.3	10.6	10.4	10.6	10.8	10.3	9.3	9.1	9.2	9.2	10.0	9.6	9.8	9.5	9.6
24	7.7	7.3	7.4	8.8	8.3	8.7	9.4	11.2	8.4	7.2	7.2	9.7	9.8	9.4	9.9	10.0	9.3	8.6	8.3	8.2	9.2	9.2	8.6	9.0	8.6	8.6
25	7.7	7.4	6.9	6.9	7.4	3.7	6.5	7.4	8.6	7.5	7.8	8.1	8.2	8.0	9.9	9.3	8.3	8.5	8.0	7.5	7.9	7.4	5.3	5.1	5.9
26	7.7	7.6	7.1	7.2	6.9	1.9	5.7	7.2	8.5	7.8	7.2	8.0	8.0	5.4	6.4	7.1	6.3	8.1	8.6	8.1	6.9	7.6	7.3	4.1	4.0	5.3
27	7.6	7.6	7.4	6.8	7.6	3.6	6.7	8.2	8.3	7.9	7.6	8.4	8.6	6.5	7.0	7.4	5.6	8.0	8.5	8.1	8.0	7.3	7.8	5.2	5.4	6.6

1984; McPherson and Smith-Lovin 1986, 1987) and in part a matter of choice (McPherson and Smith-Lovin 1987; Tuma and Hallinan 1979).

Given the overwhelming evidence of homophily with respect to numerous social dimensions, we might expect there to be homophily with respect to status, power, and expressivity in the simulated network. In fact, there is no clear tendency toward homophily with respect to all three dimensions concomitantly. The mean diagonal element is 9.96 while the mean off-diagonal element is 10.31, suggesting that these actors are slightly more likely to befriend objects with other identities than their own.

Goodness, Power, Expressivity, and Asymmetry

When measuring status and power in networks, researchers traditionally interpret asymmetric relations as evidence of the superior status of the individual on the receiving end of the esteem (e.g., Hallinan 1978-1979; Laumann 1966). Cohen (1979-1980, p. 72) remarks that homophily appears "when everyone desires contact with status superiors but settles for status equals." Are friendship choices in this simulated network asymmetric with respect to goodness and power? When the elements (e) of Table 2 are dichotomized according to the following rule,

$$e_{ij} = 1 \text{ if } e_{ij} > e_{ji}, \text{ otherwise } e_{ij} = 0, \qquad (4)$$

we would expect there to be more 1s in the lower diagonal and more 0s above, if nice people were chosen as friends by less nice people more often than the reverse. In fact, the reverse is true. Nice actors are slightly more likely to befriend less nice actors. Similarly, when the actors in Table 2 are reordered according to power and dichomitized to look for asymmetry, we find the reverse of what we might expect. Namely, powerful people are more likely to befriend weak people than are weak people to befriend powerful people. The pattern of symmetry with respect to expressivity is less clear, with a nearly equal number of 1s above and below the diagonal.[7]

Goodness, Power, Expressivity, and Emergent Structure

What else can we learn from this network of hypothetical friendships? I subjected the simulated network to some conventional clustering analyses. Burt (1978) distinguishes between graph-theoretical techniques for determining cliques based on cohesion from blockmodeling techniques based on structural equivalence. Developed for analysis of naturally occurring network data, both types of techniques rely on computer algorithms to inductively generate groupings of nodes based on some set of criteria.

Cliquing techniques identify dense regions in the network composed of groups of individuals who like (are connected to) one another. In order to analyze the

Table 3. Clique Structure of Simulated Friendship Network

Clique	Clique Members
1	4, 5, 6, 8, 12, 13, 14
2	3, 4, 5, 6, 8, 11, 13, 14, 15
3	3, 4, 5, 6, 8, 10, 13, 14, 15
4	3, 4, 5, 6, 9, 13, 14, 15
5	1, 3, 4, 5, 6, 8, 9, 14, 15
6	1, 3, 4, 5, 6, 9, 14, 15
7	5, 11, 14, 15, 16
8	5, 10, 14, 15, 16
9	5, 6, 8, 11, 13, 14, 15, 17
10	5, 6, 8, 10, 13, 14, 15, 17
11	5, 6, 8, 9, 13, 14, 15, 17
12	16
13	17
14	18
15	19
16	20
17	21
18	22
19	23
20	24, 10
21	25
22	26
23	27

clique structure of the friendship network, the likelihood matrix was dichotomized using a mean split.[8] A clique analysis was performed using UCINET IV (Borgatti, Everett, and Freeman 1994). The clique structure of this network is presented in Table 3. The most striking feature of these results is that individuals who are low in evaluation are almost always isolates. In fact, the evaluation dimension seems to be heavily involved in clique membership with people who are medium and low in evaluation being shunned and people who are high in evaluation being members of multiple close-knit friendship groups.

Results of a blockmodel analysis using Breiger's CONCOR algorithm (Breiger, Boorman, and Arabie 1975) appear in Table 4. Blockmodels identify groups of structurally equivalent individuals—those who have similar sets of relations to others—in the hopes of identifying social *positions*, or *roles*. In contrast to the clique analysis, block membership appears to depend both on evaluation and potency, with power being heavily implicated. As seen in Table 4, block 11 is composed mostly of actors who are high in power and either neutral or high in good-

Table 4a.   CONCOR Blockmodel Structure of Simulated Network

	block 11	1, 2, 3, 10, 11, 12, 20	High Potency
block 1			(High & Medium Evaluation)
	block 12	4, 5, 6, 9, 3, 14, 15	Medium Potency
			(High & Medium Evaluation)
	block 21	7, 8, 16, 17, 18, 25, 26, 27	Low Potency
			(Mixed Evaluation)
block 2			
	block 22	19, 21, 22, 23, 24	High and Medium Potency
			(Low Evaluation)

Table 4b.   CONCOR Blockmodel Structure of Simulated Network

	block 11	block 12	block 21	block 22
block 11	1	1	1	1
block 12	1	1	1	1
block 21	0	0	0	0
block 22	0	0	0	0

ness; block 12 is composed mostly of actors who are neutral in power and either high or neutral in goodness; block 21 is composed mostly of actors who are weak; and block 22 is composed mostly of actors who are mean.

What does this say about the dynamics of power and goodness in network subgroups groups? In part, it seems to provide some converging theoretical support for techniques that have been used without strong theoretical underpinnings. It is somewhat reassuring to note that, when computer algorithms are used to generate subgroupings of actors, the groupings they produce "make sense" in ways that are consistent with common usage of the procedures. Here, actors' "niceness" seems to influence clique membership and actors' power seems to influence structural equivalence. However, such "bottom up" grouping of actors relies on an assumption that is not consistent with the present design—that actors in the present matrix represent some set of individuals (or categories) that might reasonably be found in a naturally occurring group. Without this assumption, any analysis based on second-order and higher relations among the group members is not interpretable in the traditional sense.

Goodness, Power, Expressivity, and Friendship Choice

Fortunately, the design of the simulated matrix enables us to make a priori groupings of actors based on the social dimensions that affect control theory assumes to be most important. By design, the likelihood matrix in Table 2 is already "blocked" according to evaluation. Table 5 shows a likelihood matrix by that blocking, giving us a highly reduced summary of how evaluation influences friendship choice.

According to this summary, homophily seems to operate among good and neutral identities, but not among bad identities. In fact, actors operating in bad identities cannot "befriend" anyone and maintain their own negative associations. Actors who are neutral in evaluation are relatively likely to befriend anyone.

Nice actors can befriend other nice actors and neutral-evaluation actors without doing damage to their identities.[9] While moderately symmetric in this reduced form, we see the direction of the asymmetry described previously—"good" actors are more likely to befriend "bad" actors than vice versa. In fact bad or mean actors are also more likely to befriend other mean actors than to befriend good or nice actors. The greatest likelihoods are for friendships among nice actors.

Table 6 displays the reduced likelihood matrix when actor order is rearranged and then blocked by potency. The most notable aspect of this image is the *lack of homophily among high power actors*. In fact, according to this summary, powerful actors will sooner befriend weak others than other powerful actors. Like neutral evaluation actors, neutral power actors seem able to befriend actors all over the potency dimension without disrupting their own identity-affect. Weak actors have difficulty befriending anyone, most of all other weak actors.

DISCUSSION

After examining associates labeled as "friends" by a large cross-sectional survey of adults, Fischer (1982) concluded that, compared to other close relations, we use the term *friend* to describe relations who are non-kin, who we've known a long time, with whom we have social (rather than intimate or material) relations, and with whom we do not have institutionally clear role-relations. According to the

Table 5.   Reduced Likelihoood Matrix Blocked by Evaluation

	Good (1,2,3,4, 5,6,7,8,9)	Neutral (10,11,12,13, 14,15,16,17,18)	Bad (19,20,21,22, 23,24,25,26,27)
Good	12.24	11.93	9.78
Neutral	11.45	11.75	10.50
Bad	7.99	8.78	8.29

Table 6. Reduced Likelihoood Matrix Blocked by Potency

	Powerful (1,2,3,10,11, 12,19,20,21)	Neutral (4,5,6,13,14, 15,22,23,24)	Weak (7,8,9,16,17, 18,25,26,27)
Powerful	9.80	11.30	11.63
Neutral	10.42	11.32	11.29
Weak	9.66	9.47	7.84

Table 7. Reduced Likelihoood Matrix Blocked by Activity

	Lively (1,4,7,10,13, 16,19,22,25)	Neutral (2,5,8,11,14, 17,20,23,26)	Passive (3,6,9,12,15, 18,21,24,27)
Lively	9.66	10.09	10.03
Neutral	10.64	10.91	10.70
Passive	10.23	10.40	10.07

INTERACT dictionary, to "befriend" someone is a good, powerful, though not-so-lively act (1.56, 1.63, -0.4). We see the implication of these meanings in the findings that very bad, very weak, or very lively people cannot "befriend" others without doing some damage to their own identity-affect. The act of befriending (deliberately) does not capture the full repertoire of behaviors that could lead to a "network tie" in the sense that it is often used. Other behaviors, "to confide in," "to ally with," "to talk with," "to support," would be expected to yield somewhat different results. While not capturing the full complexities and ambiguities of naturally occuring friendships, the advantage of simulations in this context is the ability to focus on one specific form of relation—that based on the good, powerful act of "befriending."

This simulated network consisted of an artificial community—indeed, it contained some members who could not be sampled from any naturalistic population! Any results that take into account structure of relations to others (like cliquing and blockmodeling) are not interpretable in a traditional sense without making the unlikely assumption that these individuals might make up either some naturally occurring group or be comparable to some meaningful sample of individuals from a population. However, even results that take into account such higher-order relations are interpretable if we compare across goodness, power, and expressivity as varied in the design. So, while we would not put much stock into measures of centrality in this hypothetical matrix, we can find something of interest in the observations that bad actors did not find their way into emergent cliques and that potency had a large impact on the structural equivalence of actors in the group.

Our identities locate us in social space through relationships and memberships (Stone 1962). Sociologists have attended to the conception that identities and identity-based action derive from social location. However, our theories also suggest that these locations are, in part, products of our identities. Affect control theory's generative, cybernetic model of social interaction allows us to consider the interactive and processual nature of the relationship between self and society.

While only a tentative theoretical exploration, hopefully, this work deepens our understanding of the role of power and status in structuring human interaction. The logic of these simulations suggests a potentially important extension to affect control theory—expanding our ability to model the role of affective dynamics in shaping both our social actions and, consequently, our social environments. If the theory's assumptions are valid and its empirical base well-grounded, then theory based simulations of friendship likelihoods should produce intuitive patterns of results in the form of testable predictions of the theory.

Indeed this simulated friendship network did produce, in part, findings that are consistent with intuitions and existing empirical evidence. The friendship likelihoods revealed a tendency toward homophily respect to goodness, cohesion-based subgroups that seemed related to goodness, and position-based subgroups that appeared related to power. In addition, however, the simulations produced some less intuitive, but testable predictions: (1) powerful people are relatviely unlikely to befriend other powerful people, and (2) nice people more likely to befriend mean people than vice versa. These predictions require both testing and reconciliation with existing claims (theoretical or empirical).

ACKNOWLEDGMENTS

I would like to thank Barry Markovsky and Johnmarshall Reeve for helpful comments on this manuscript and James W. Balkwell, Lynn Smith-Lovin, and J. Miller McPherson for contributions during discussions of this material.

NOTES

1. This profile comes from the INTERACT U.S. (male) dictionary, available in Heise and Lewis (1988). For the United States, Smith-Lovin and Heise (1988) have compiled EPA ratings for a corpus of 345 social settings, 765 identities, 600 social behaviors, and 440 emotions/person modifiers. In this dictionary, EPA profiles for each word are based on the mean ratings of approximately 25 male and 25 female undergraduates at a southern U.S. university. This dictionary contains the EPA profile of concepts outside the context of social events. These profiles form the affective points of reference that affect control theory terms *fundamental sentiments*.

2. INTERACT 2 also contains separate equation sets and dictionaries for Germany, Japan, and Canada. This software is available through direct correspondence with David R. Heise, Department of Sociology, University of Indiana, Bloomington, Indiana.

3. The $B_e O_e$ interaction also implies that one can qualify the negative effects of a negative behavior by directing it toward a mean object. In contrast, one can intensify the negative effects of committing a harmful behavior by directing it toward a nice person.

4. What constitutes acceptable levels of deflection is not clear from current formulations of affect control theory. In this, and previous work, I sidestep this problem in part by considering comparative deflections.

5. While evaluation, potency, and activity have been demonstrated to be fundamental dimensions of meaning in 21 different language communities worldwide (Osgood, May, and Miron 1975), affect control theory assumes that the EPA measures for particular stimuli to vary across culture. Within culture (or subculture), EPA measures for the various components of social interactions (social identities, interpersonal acts, environmental settings, etc.) are assumed to be general across different individuals.

6. Simulated actors also befriended actors who shared their identity profiles (because I consider these role-identities rather than individuals).

7. Specific values for these analyses can be obtained by applying the rule to various arrangements of Table 2 or directly from the author.

8. The choice of the mean cutoff was arbitrary. Other reasonable cutoff values yielded similar results. When the cutoff is higher, however, the number of cliques with more than one member decreases.

9. Although I am discussing deflection in terms of disruption to identity-meanings, it is important to keep in mind that deflection is an interaction-based term—it reflects meaning disruption in all elements of the social event (actor, object, behavior, setting). In the reported simulations all elements other than actor identity and object identity are constant. Therefore, it is appropriate to interpret deflection comparisons as reflecting differing amounts of disruptions in the identities of the actor and object.

REFERENCES

Backman, C. W., and P. F. Secord. 1962. "Liking, Selective Interaction, and Misperception in Congruent Interpersonal Relations." *Sociometry* 25: 321-335.
Bem, D. J. 1972. "Self-Perception Theory." In *Advances in Experimental Social Psychology*, edited by L. Berkowitz. New York: Academic Press.
Borgatti, S., M. Everett, and L. Freeman. 1994. *UCINET IV.* Version 1.35. Columbia: Analytic Technologies.
Berscheid, E., and E. Walster. 1978. *Interpersonal Attraction*. Reading, MA: Addison-Wesley.
Breiger, R. L., S. A. Boorman, and P. Arabie. 1975. "An Algorithm for Clustering Relational Data With Application to Social Network Analysis and Comparison With Multidimensional Scaling." *Journal of Mathematical Psychology* 12: 328-383.
Burke, P. J. 1980. "The Self: Measurement Requirements From an Interactionist Perspective." *Social Psychology Quarterly* 43: 18-29.
Burke, P. J., and D. Reitzes. 1981. "The Link Between Identity and Role Performance." *Social Psychology Quarterly* 44: 83-92.
_____. 1991. "An Identity Theory Approach to Commitment." *Social Psychology Quarterly* 54: 239-251.
Burt, R. S. 1978. "Cohesion Versus Structural Equivalence as a Basis for Network Subgroups." *Sociological Methods and Research* 7: 189-212.
Campbell, K. E., P. V. Marsden, and J. S. Hurlbert. 1986. "Social Resources and Socioeconomic Status." *Social Networks* 8: 97-117.
Carley, K. M. 1986. "Knowledge Acquisition as a Social Phenomenon." *Instructional Science* 14: 381-438.
Cohen, J. 1979-1980. "Socioeconomic Status and High School Friendship Choice: Elmtown's Youth Revisited." *Social Networks* 2: 65-74.
Cooley, C. H. [1902].1956. *Social Organization and Human Nature and the Social Order*. New York: Free Press.

Erickson, B. H. 1988. "The Relational Basis of Attitudes." In *Social Structures: A Network Approach*, edited by B. Wellman and S.D. Berkowitz. New York: Cambridge University Press.

Feld, S. L. 1981. "The Focused Organization of Social Ties." *American Journal of Sociology* 86: 1015-1035.

_____. 1982. "Structural Determinants of Similarity Among Associates." *American Sociological Review* 47: 797-801.

_____. 1984. "The Structured Use of Personal Associates." *Social Forces* 62: 640-52.

Festinger, L., S. Schachter, and K. Back. 1950. *Social Pressures in Informal Groups: A Study of Human Factors in a Housing Community.* New York: Harper.

Fischer, C. S. 1982. "What Do We Mean by 'Friend'? An Inductive Study." *Social Networks* 3: 287-306.

Hallinan, M. T. 1974. *The Structure of Positive Sentiment.* New York: Elsevier.

_____. 1978-1979. "The Process of Friendship Formation." *Social Networks* 1: 193-210.

Hays, R. B. 1984. "The Development and Maintenance of Friendship." *Journal of Personality and Social Psychology* 1: 75-98.

Heise, D. R. 1979. *Understanding Events: Affect and the Construction of Social Action.* New York: Cambridge University Press.

_____. 1987. "Affect Control Theory: Concepts and Model." *Journal of Mathematical Sociology* 13: 1-33.

_____. 1991. *Interact 2: A Computer Program for Studying Cultural Meanings and Social Interactions.* Bloomington: Department of Sociology, Indiana University.

_____. 1992. *Affect Control Theory's Mathematical Model, With a List of Testable Hypotheses: A Working Paper for ACT Researchers.* Bloomington: Department of Sociology, Indiana University.

Heise, D. R., and E. Lewis. 1988. *Introduction to INTERACT.* Raleigh, NC: National Collegiate Software Clearinghouse, North Carolina State University.

Hogg, M. A. 1992. *The Psychology of Group Cohesiveness: From Attraction to Social Identity.* New York: Harvester Wheatsheaf.

Johnsen, E. C. 1986. "Structure and Process: Agreement Models for Friendship Formation." *Social Networks* 8: 257-306.

Kemper, T. D. 1978. *A Social Interactional Theory of Emotions.* New York: Wiley.

Kemper, T. D., and R. Collins. 1990. "Dimensions of Microinteraction." *American Journal of Sociology* 96: 32-68.

Laumann, E. O. 1966. *Prestige and Association in an Urban Community.* Indianapolis: Bobbs Merrill.

Lazarsfeld, P. F., and R. K. Merton. 1954. "Friendship as a Social Process." In *Freedom and Control in Modern Society.* New York: Litton.

MacKinnon, N. J. 1994. *Symbolic Interactionism as Affect Control.* Albany, NY: SUNY Press.

MacKinnon, N. J., and D. R. Heise. 1987. "Affective Bases of Likelihood Judgments." *Journal of Mathematical Sociology*, 13: 133-151.

_____. 1993. "Affect Control Theory: Delineation and History." In *Theoretical Research Programs: Studies in the Growth of Theory*, edited by J. Berger and M. Zelditch, Jr. Stanford, CA: Stanford University Press.

Markovsky, B., D. Willer, and T. Patton. 1988. "Power Relations in Exchange Networks." *American Sociology Review* 53: 220-36.

Marsden, P. 1988. "Homogeneity in Confiding Relationships." *Social Networks* 10: 57-76.

McPherson, M. J., P. Popielarz, and S. Drobnic. 1991. "Social Networks and Organizational Dynamics." Paper presented at the annual meetings of the American Sociological Association.

McPherson, M. J., and L. Smith-Lovin. 1986. "Sex Segregation in Voluntary Associations." *American Sociological Review* 51: 61-80.

_____. 1987. "Homophily in Voluntary Organizations: Status Distance and the Composition of Face-to-Face Groups." *American Sociological Review* 52: 370-379

Mead, G. H. 1934. *Mind, Self, and Society.* Chicago: University of Chicago Press.

Newcomb, T. M. 1961. *The Acquaintance Process.* New York: Holt, Rihehart, and Winston.

Newcomb, T. M. 1981. "Heiderian Balance as a Group Phenomenon." *Journal of Personality and Social Psychology* 40: 462-867.

Osgood, C. E., W. H. May, and M. S. Miron. 1975. *Cross-Cultural Universals of Cultural Meaning.* Urbana: University of Illinois Press.

Osgood, C. E., G. C. Suci, and P. H. Tannenbaum. 1957. *The Measurement of Meaning.* Urbana: University of Illinois Press.

Powers, W. T. 1973. *Behavior: The Control of Perception.* Chicago: Aldine.

Ridgeway, C., and H. Walker. 1995. "Status, Authority and Legitimacy." In *Sociological Perspectives on Social Psychology,* edited by K. Cook, G. Fine, and J. House. Boston: Allyn and Bacon.

Robinson, D. T. 1992. *Emotions and Identity Negotiation: The Effects of Nonverbal Displays on Social Inference and Selective Interaction.* Doctoral dissertation. Department of Sociology, Cornell University.

Robinson, D. T., and L. Smith-Lovin. 1992. "Selective Interaction as a Strategy for Identity Maintenance: An Affect Control Model." *Social Psychology Quarterly* 55(1): 12-27.

Rosenberg, M. 1979. *Conceiving the Self.* New York: Basic.

Serpe, R. T. 1991. "The Cerebral Self: Thinking and Planning about Identity-Relevant Activity." Pp. 55-74 in *The Self-Society Dynamic,* edited by J. A. Howard and P. L Callero. New York: Cambridge University Press.

Serpe, R. T., and S. Stryker. 1987. "The Construction of Self and the Reconstruction of Social Relationships." In *Advances in Group Processes,* Vol. 4, edited by E. J. Lawler and B. Markovsky. Greenwich, CT: JAI Press.

Skvoretz, J., and D. Willer. 1991. "Power in Exchange Networks: Setting and Structure Variations." *Social Psychology Quarterly* 54: 224-38.

Smith-Lovin, L. 1987. "Impressions From Events." *Journal of Mathematical Sociology* 13: 35-70.

_____. 1990. "Emotion as the Confirmation and Disconfirmation of Identity: An Affect Control Model." In *Research Agendas in Emotion,* edited by T. Kemper. Albany, NY: SUNY Press.

Smith-Lovin, L., and D. R. Heise. 1988. *Analyzing Social Interaction: Advances in Affect Control Theory.* New York: Gordon and Breach Science Publishers.

Stone, G. P. 1962. "Appearances and the Self." In *Human Behavior and Processes,* edited by A. M. Rose. Boston: Houghton Mifflin.

Stryker, S. 1981. *Symbolic Interactionism: A Social Structural Version.* Menlo Park, CA: Benjamin-Cummings.

Suitor, J. J. 1987. "Friendship Networks in Transitions: Married Mothers Return to School." *Journal of Social and Personal Relationships* 4: 445-461.

Suitor, J. J., and S. Keeton. In Press. "Once a Friend, Always a Friend? Effects of Homophily on Women's Support Networks Across a Decade." *Social Networks.*

Suitor, J. J., K. Pillemer, and S. Keeton. 1995. "When Experience Counts: The Effects of Experiential and Structural Similarity on Patterns of Support and Interpersonal Stress." *Social Forces* 73:1573-1588.

Swann, W. B., Jr. 1985. "The Self as Architect of Reality." Pp. 100-125 in *The Self and Social Life,* edited by B. R. Schlenker. New York: McGraw Hill.

_____. 1987. Identity Negotiation: Where Two Roads Meet. *Journal of Personality and Social Psychology* 53: 1038-1051.

Tajfel, H. 1981. *Human Groups and Social Categories: Studies in Social Psychology.* New York: Cambridge University Press.

_____. 1982. *Social Identity and Intergroup Relations.* New York: Cambridge University Press.

Taylor, S. E., and J. D. Brown. 1988. "Illusion and Well Being: Some Social Psychological Contributions to a Theory of Mental Health." *Psychological Bulletin* 103:193-210.

Tesser, A., C. J. Pilkington, and W. D. McIntosh. 1989. "Self-Evaluation Maintenance and the Mediational Role of Emotion: The Perception of Friends and Strangers." *Journal of Personality and Social Psychology* 57(3): 442.

Tuma, N. B., and M. T. Hallinan. 1979. "The Effects of Sex, Race, and Achievement on Schoolchildren's Friendships." *Social Forces* 57: 1286-1309.

Verbrugge, L. M. 1977. "The Structure of Adult Friendship Choices." *Social Forces* 56: 576-597.

_____. 1979. "Multiplexity in Adult Friendships." *Social Forces* 57: 1286-1309.

_____. 1983. "A Research Note on Adult Friendship Contact: A Dyadic Perspective." *Social Forces* 62: 78-83.

IDENTITY WORK AS GROUP PROCESS

Michael L. Schwalbe and Douglas Mason-Schrock

ABSTRACT

This paper examines the process whereby groups create the signs, codes, rites of affirmation, and boundaries upon which the existence and maintenance of shared identities depend. We call this process subcultural identity work, arguing that it consists of four essential parts: defining, coding, affirming, and policing. Two empirical illustrations of subcultural identity work are offered—one derived from a study of men who used myth and poetry to remake "man" as a moral identity, another derived from a study of transsexuals. Other studies of identity work are reviewed and discussed. The concepts of oppressive identity work and oppositional identity work are developed as tools for understanding the identity struggles that are often at the core of intergroup conflict.

IDENTITY WORK AS GROUP PROCESS

Social psychologists are losing their grip on identity. A concept that was once our stock-in-trade is now widely used by literary theorists, historians, and cultural studies scholars. This might seem like success except that much of the new writing on identity takes little heed of what social psychologists have had to say about the matter. Disciplinary myopia is one reason for this. Another reason is a failure on

Advances in Group Processes, Volume 13, pages 113-147.
ISBN: 0-7623-0005-1

our part to move our thinking about identity beyond the individual level, as Anselm Strauss (1959, p. 175) once urged us to do: "A social psychology without full focus upon history is a blind psychology. A concern with personal styles, strategies, careers—in short, with personal identities—requires a serious parallel concern with shared, or collective identities, viewed through time." In over 30 years Strauss's challenge has barely been met.

In American social psychology, identities typically have been studied as parts of the self-concept, the goals being to find out what these identities mean, how they are organized, and how they affect behavior. Most studies have used questionnaires to elicit people's reports about the identities they claim for themselves. This line of inquiry seeks to understand identities as the psychic property of individuals. Another line of inquiry, which we pursue here, in keeping with Strauss's admonition, aims to understand how identities are created, used, and changed in interaction. Our interest, in other words, is in identity making as something that people do together.

This view of identity as a joint accomplishment is a mainstay of symbolic interactionism (Blumer 1969; Hewitt 1989; McCall and Simmons 1978; Strauss 1959), but it is most closely associated with the dramaturgical perspectives of Burke (1957) and Goffman (1959). In this view, social life is made up of connected dramatic enactments through which people communicate their dispositions and coordinate action (for an overview, see Brissett and Edgley 1990, pp. 1-46). It is in and through these dramatizations that selves are signified and affirmed, both to others and reflexively to one's self. The goal of dramaturgical analysis is, then, to describe and explain how people construct their performances and thereby create themselves and each other as social objects.

Work in this tradition, broadly construed, has examined how identities are created, negotiated, and repaired through talk (Hadden and Lester 1978; Hewitt and Stokes 1975; Lyman and Scott 1968; Snow and Anderson 1987); how dress is used to announce identities (Davis 1992; Finkelstein 1991; Stone 1981); how people "do gender" (Kessler and McKenna 1978; West and Zimmerman 1987); how identities are generated through story telling (Gergen and Gergen 1983; Young 1989); how identities are asserted through deviant acts (Lofland 1969); how identity boundaries are policed (Kitzinger 1989; Nagel 1994; Walker 1993;); how social movement activists create collective identities (Friedman and McAdam 1992; Hunt and Benford 1994; Taylor and Whittier 1992); and how selves are created with signs (Perinbanayagam 1991; Schwalbe 1993; Singer 1980). Psychologists who have done experiments to test Goffman's "intuitions" about impression management are also part of the tradition (see, e.g., Baumeister 1982; Jones and Pittman 1982; Wicklund and Gollwitzer 1982; Schlenker 1980; Tedeschi 1981). The common thread is a concern with the process of identity-making, rather than with identities as parts of the self-concept.

Most of this previous research and theorizing has treated identity work as an activity of individuals. In this paper our focus is on identity work as a group process. We argue that to understand identity-making it is necessary to examine not only individual self-presentations but the joint creation of the symbolic resources upon which those presentations depend—an activity we refer to as subcultural identity work. We begin with an overview of the literature on identity work. We then develop a theoretical sketch of this activity, drawing on our own and other recent studies for empirical illustrations. In the final section of the paper, we introduce the concepts of oppressive and oppositional identity work to suggest how identity-making can be understood as part of a process of cultural struggle.

IDENTITY WORK

A great many notions lurk behind the term "identity," and we don't intend to deal with them all here (Weigert, Teitge, and Teitge 1986 have provided this service). We will focus instead on what has come to be called *identity work*, which is anything people do, individually or collectively, to give meaning to themselves or others. Identity work is thus largely a matter of signifying, labeling, and defining. It also includes creation of the codes that enable self-signifying and the interpretation of others' signifying behavior. As we will discuss later, our definition of identity work overcomes the individualistic bias present in other treatments of the topic.

A focus on identity work suggests a somewhat different view of identity than that found in the mainstream of social psychology, where identities are seen as meanings given to the self, that is, as parts of the self-concept (Burke and Reitzes 1981; Gordon 1968; Stryker 1968). It is important, of course, to recognize that people apply labels to themselves, and that these labels evoke both social and internal responses, thus giving the individual meaning as a social object. But we prefer to avoid saying that identities are "meanings given to the self," as this bit of shorthand is misleading. Identities, as we see them, are *indexes of the self*. By this we mean that identities are signs that refer to qualities of the identity claimant. An identity, then, is not a meaning but *a sign that evokes meaning*, in the form of a response aroused in the person who interprets it. What identities signify, more specifically, are the powers, status, inclinations, and feelings—in short, the self—of the persons to whom they attach. In this view, an identity is not a constituent feature of the self but a semiotic tool for situated self making.

There are two kinds of "work" implied by this view of identity. One kind is necessarily communal: the creation of identities as widely understood signs with a set of rules and conventions for their use. Another kind is individual, though not private: the use of these signs, rules, and conventions by individuals to create images of themselves in interaction. It is through these kinds of work that meanings are made in social life, specifically, the meanings that attach to persons as social objects and by which the responses of others are mobilized (Stone 1981).

To call this meaning-making activity *work* is to call attention to it as involving purposes, strategies, and, sometimes, the overcoming of resistance. To call it work also makes it more visible as an object of analysis. Goffman (1959, 1961, 1963, 1967) provided the paradigm examples of this kind of analysis. He exposed a variety of techniques that people use to make identity claims, to create creditable images of themselves, and to repair damaged selves. Although he sometimes referred to interaction as an "information game," he also referred to what people do in interaction as work of various kinds (e.g., "face work," "remedial work"), emphasizing the care and skill that such activity requires and its fateful consequences.

Using the language of work to describe what goes on in interaction also reminds us that the making of selves and social life are accomplishments. Interaction doesn't just happen; nor are people compelled to do what they do by norms, values, expectations, and rules. Rather, in Goffman's (1983) view, the rules and conventions that constitute the interaction order are resources that people can use to elicit responses from each other and thereby coordinate action, while also protecting the feelings attached to selves. Moreover, this is always collaborative work, in that the interaction order must be jointly upheld to remain useful, and we must help others establish and sustain their virtual selves in every encounter.

Implicit in this view of interaction as an accomplishment is the idea that, while certain kinds of episodes can be routinized, they are never entirely unproblematic. Even when familiar identities are claimed or imposed, there is always room for negotiation. It might be clear, for example, that one is claiming the identity "professor" in a given situation; yet the task remains to signify exactly what *kind* of professor one is going to be. And there is always room for things to go wrong. Discrepant signs can appear and undermine an initial identity claim. Efforts to manage these contingencies—that is, to make sure the self is signified in a way that suits an unfolding situation—are part of the identity work inescapably demanded by social life.[1]

Analyses of Rhetoric and Talk

Various analysts, following Goffman, have sought to show how people collaborate in making and upholding identity claims. Lyman and Scott's (1968) analysis of accounts is, at one level, an analysis of rhetorical strategies used to do identity work. As they note, giving and honoring accounts is a "manifestation of the underlying negotiation of identities." Excuses and justifications of various kinds, as well as other disclaimers (Hewitt and Stokes 1975), can be seen as forms of talk used to assert, protect, and repair situated identities. "Mere talk" is still work because accounts must be fashioned carefully, in the appropriate idiom, to be effective. An audience must also collaborate by honoring the accounts given.

Talk has been a main focus in studies of identity work. Hadden and Lester (1978), for example, describe a set of "generic practices for proffering [verbal] accounts of self" (p. 335). Their analysis focused on the practice of "locating," by which is meant the practices of (a) linking one's self to a network of similar others, and (b) depicting the self as different from others (1978, pp. 338-339). What the practice of locating does is to assemble verbal identity markers that establish an identity in the present. The practices of retrospecting—linking identity claims to biography—and prospecting—linking identity claims to an anticipated future— are said to aid locating. Hadden and Lester also suggest that identity work involves giving not only information about one's self but, along with that information, a scheme for interpreting it.

In a similar vein, Snow and Anderson (1987) describe three patterns of identity talk among the homeless: distancing, embracement, and fictive storytelling. Snow and Anderson link the use of these forms of identity talk to length of time on the street, showing, for example, that "identity statements implying choice [about being homeless] can best be regarded as manifestations of life on the street rather than indicators of initial precipitants" (p. 1364). They also suggest that storytelling flourishes because the homeless know little about each other's biographies, they have nothing to gain and violence to fear if they challenge dubious identity claims, and they need each other's help in doing face work to cope with the indignities they suffer. Snow and Anderson's study thus goes beyond Hadden and Lester by showing how forms of identity talk are adaptive responses to the material conditions of people's lives.

Identity talk may also be linked to the political projects in which people are involved, as Hunt and Benford (1994) show in their analysis of participants in the peace and justice movement. Participants in this movement communicated their identities as activists by telling stories about becoming "aware, active, committed, and weary." Through telling such stories, Hunt and Benford show activists not only stake out their personal identities but also define the political playing field and the other actors (individuals and groups) in it. As Hunt and Benford argue elsewhere (Hunt, Benford, and Snow 1994), much identity talk in social movements serves this latter function of defining the character of movement organizations. We will say more about this when we distinguish between individual, collective, and subcultural identity work.

Toward a Wider Semiotic View

The studies cited above show the importance of talk for the creation of identities in face-to-face encounters. What's more, they show how people adopt and adapt various rhetorical strategies for presenting themselves as they want to be seen in any given situation. It seems there is a vast range of strategies and forms of rhetoric that different people use under different conditions. There are perhaps as many different ways of "talking identity" as there are groups of people who share patterns of talking. By studying different groups and subcultures, one might build a long list of strategies. It seems reasonable, however, to wonder how an analysis of identity work might go beyond the cataloging of forms of self-presentational rhetoric.

One avenue is to conceive of identity work more broadly, as involving not just talk but all the ways in which the self is signified and identities are claimed.

Stone's (1981) analysis of appearance and the self is a classic in this genre. Stone did not merely say that people use clothes to make identity claims, but that clothes, as parts of our *programs* of appearance, serve to announce identities, show values, express moods, and propose attitudes (i.e., tendencies to act). Appearance, in other words, is programmed—organized in accord with a local code—to signify many things at once. To mobilize the responses of others as we would like, we must learn to put the pieces of the program together just right. "Costume," Stone said, "is a kind of magical instrument." But it is an instrument of mundane origin, arising out of skillful dramaturgical work.

Interactionists have analyzed dress from various angles. In some cases the analysis is simply a decoding of the otherwise cryptic styles of subcultural groups, but usually there is an attempt to link these styles to larger themes and problems in the culture. For example, Young (1990, pp. 339-356) draws connections between the lack of outlets for political participation in U.S. society and the reliance on T-shirt sloganeering as a way to cheaply and safely signify one's political sensibilities. Fred Davis (1992) shows how fashion designers exploit the identity ambivalences that are endemic to complex, mass societies. New designs are marketed and absorbed, Davis argues, because people constantly seek to signify themselves in a way that resolves these identity ambivalences and tensions.

Joanne Finkelstein (1991) adds a facet to the analysis of fashion, arguing that the skillful use of fashion is itself an important signifier. As she puts it, "the value of one's physical appearance rests upon its being interpreted as a measure of one's competence with the tools and devices which are available to transform and style the human body" (p. 183). Her point is that displaying the *ability* to make one's self into a powerful evocative object is itself a form of identity work. Finkelstein's analysis thus broadens our conception of how identity work is accomplished when self-presentations are crafted. She shows that meanings are made in several ways at once.

Analyses, like Finkelstein's, move in the direction of semiotics.[2] Here the question about identity work becomes: How do people *signify the self?* An answer to this question goes beyond talk and clothes to include all the ways in which signs of the self are fabricated and wielded. Such signs can now be understood to include not only situated talk and appearance but also complex acts that extend over time. These acts might include choices of consumer goods; the choice and shaping of one's ego extensions; choices of sports activities, political alignments, friends, books, music, theoretical perspectives, research projects, and so on. Any act that reflects a desire to signify one's qualities to others, and thereby define one's self as a social object, can be considered an instance of identity work.

The semiotic turn raises other important questions: How do people establish connections between signs and themselves? How do people make the signs that do identity work for them? How do people ensure that such signs will work as intended? By what logic or grammar are discrete signs assembled into programs of self-signification? In the view implied by these questions, identity work is seen as a kind of *semiotic bricolage*, wherein people create themselves as social objects by fabricating signs of the self out of whatever resources they can find.[3] Though, if necessary, people will also create the tools—the signs, codes, and rites of affir-mation—that allow them to make of themselves what they would like to be.

Robert Perinbanayagam (1985, 1991) has theorized extensively about these matters, exploring the relationship between signing (i.e., signifying behavior) and the self.[4] His argument, in brief, is that through various kinds of signifying acts:

> ... the self and its intentionalities [are] objectified, so that the other may attend to them. In creating these discourses, a human constructs acts directed toward others that embody his or her own intentions—indeed his or her very self. In taking the other's role we take and interpret these discursive texts that do not merely provide pleasure but are the very matter by which we manage to live with others. In sum, the intentionality of the other is manifest in the details that constitute discursive texts, details manifested as signs. (1991, p. 8)

Although Perinbanayagam refers to "discursive texts," his larger analysis takes into account the many ways that intentionalities are objectified in the form of signs. What this perspective offers is a still broader view of the forms identity work can take and what it accomplishes. In short, any word, gesture, or deed that objec-tifies our intentionality—any act that is offered or seen as an expression of our sub-jectivity—can be an instance of identity work. In social life such acts are read as signs of the impulses that make us what we are.

This perspective lets us see identity work as including all attempts—whether obvious or subtle, clear or ambiguous—to signify one's subjectivity. In other words, identity work includes not just signification of one's social location but one's powers, worth, inclinations, and feelings. We can thus ask how any act might be intended, perhaps only dimly or allusively, to signify qualities of the self. We can also ask how it is that people create and use a system of signs for expressing who and what they are. As we will propose later, the process sometimes involves creating special social representations—moral identities—that can serve as uni-versal indexes of virtuous selfhood.

We have so far skirted the question of whether acts of self-signification must be intentional to be considered instances of identity work. Our answer is yes and no—yes, in that *at one time* such acts must have been intentional, and no, in that identity work can become a matter of habit and slip beneath conscious awareness. For example, a teenager might studiously choose which brand of cigarette to smoke to signify a particular kind of self. But once the habit and brand loyalty are established, there may be little awareness of this choice as an identity statement. The same point would apply to the clothes we wear. What is at first a matter of choice—"Is this the kind of jacket a person like me would be seen wearing?"—

later becomes a matter of habit: "This is simply the jacket I wear; I don't even think about it." It thus seems likely that people will tend to forget, if only as a result of cognitive economizing, how many of their habits originated as acts of self-signification. By implication, people will also tend to forget how much identity work they are doing at most times.

There is other literature that can be construed as dealing with identity work, if not in exactly those terms. The same basic ideas that we have extracted in this section run through all this literature: (a) people use a variety of rhetorical strategies to assert, protect, and repair images of themselves in interaction; and (b) these images, or situated identities, are always the results of negotiation. These are important insights, but as guides for analysis they have led to little more than a cataloging of strategies. We think the semiotic view opens up greater analytic possibilities (cf. Manning and Cullum-Swan 1994) since it leads us to ask not only about rhetorical strategies but also about the joint creation of the symbolic resources that make those strategies possible and effective in the first place.

Individual, Collective, and Subcultural Identity Work

Most of the research and theorizing discussed above is concerned with what can be called individual identity work. This individualistic bias is reflected in Snow and Anderson's (1987) definition of identity work as:

> The range of activities individuals engage in to create, present, and sustain personal identities that are congruent with and supportive of the self-concept ... So defined, identity work may involve a number of complementary activities: (a) procurement or arrangement of physical settings or props; (b) cosmetic face work or the arrangement of personal appearance; (c) selective association with other individuals and groups; and (d) verbal construction and assertion of personal identities. (p. 1348)

Note that the emphasis in this definition is on what individuals do to signify their *personal* identities. We have been pursuing a broader view of identity work as including anything people do, individually or collectively, to give meaning to themselves or others. In this section we want to more explicitly distinguish between the identity work done by individuals and that done collaboratively by people who share an identity.

To call what individuals do *individual* identity work is not to say that it needs no audience or that the signs it involves are not jointly created. Identity work is always a social activity, in one way or another. But identity work can be done in ways that primarily serve either the peculiar needs of individuals or the common needs of people in a group. Acts that are tailored to signify the subjectivity of one person, whether these acts are performed in a single encounter or over time, are instances of individual identity work.

Going beyond the individual level, social movements scholars have studied what they call "collective identity," which refers to the meaning given to a political group or social movement organization (Hunt, Benford, and Snow 1994; Taylor and Whittier 1992). A collective identity, in this view, is the identity of *a corporate body*. In the social movements literature, "creating a collective identity" is thus seen as a matter of people in social movement organizations defining a political situation and the character of contending parties within it.[5]

Hunt and Benford's (1994) study of identity talk in the peace and justice movement, for example, examined the making of both individual and collective identities. As they saw it, "identity talk" included not only stories activists told about themselves but also their talk about the struggle in which they were engaged, about the character of their organization and the character of other groups in the same political arena, and even about outsiders. While this analysis turns up what are no doubt key aspects of group process in social movement organizations, it fails to distinguish between: (1) the work people do to signify their personal identities, (2) the resource-creating work done in groups by people who share an identity, and (3) the ideological work that political actors do to define situations to their advantage. All these efforts at making meanings can overlap or align, but they are not the same thing.

Our primary interest is in the second type of identity work noted above, that is, the creation of symbolic resources by people who share an identity. Hunt and Benford (1994) allude to this kind of identity work when they describe movement leaders giving pep talks to the rank and file: "Typically, a representative of an SMO provides a sketch of the organization's collective identity and indicates what kind of people associate with the group" (p. 496). In this case one person takes the lead in creating not only an image of the organization but, more importantly, an image of the kind of people who belong to such an organization. This is an example of real identity-making: leaders telling members—who share an identity as members—what it is that their belonging signifies. The shared identity of, say, "Nation of Islam member" is thus invested with meaning and can be used as a sign of the self.

The key distinction for our purposes is between the dramaturgical work people do to signify who and what they are as individuals and the work people do together to create the signs, codes, and rites of affirmation that become shared resources for identity-making. Our focus is on the latter, the group process, which we call subcultural identity work. We call it *subcultural* rather than collective identity work to avoid confusion with the notion of collective identity found in the social movements literature.[6] To call it subcultural also takes us beyond the realm of social movements to include any group wherein people create or redefine an identity to which they all lay claim. *Sub*cultural further suggests, more accurately, the creation of a body of meanings, signs, and signifying practices that are distinct from, yet linked to, a larger culture. A further sociological point is implied by the emphasis on a subcultural, rather than simply a group, process: identity-making is simultaneously a making of culture (Fine and Kleinman 1979; Nagel 1994).

Before we outline the process of subcultural identity work, we want to briefly remark on why people do identity work. This is a departure from the usual dramaturgical disregard for *why* in favor of *how*. While our primary interest is indeed in how identity work is done, we think that an analysis of how always rests on assumptions about why and that it is best to make these assumptions explicit. We also think that a social psychology focused on identity work is valuable for seeing connections between motivation and behavior on the one hand and culture and structure on the other. To see these connections, some consideration of the why question is necessary.

Identity work is, first of all, a requirement of social life. We must coordinate action with others, and we cannot do this without being able to recognize others as individuals, or at least as types of individuals. In face-to-face encounters we must know roughly what to expect from whom, just as they must know roughly what to expect from us. It is by doing identity work that we make ourselves functionally recognizable; it is also how we signify our intentions, abilities, and expectations so that others can adjust their behavior toward us. There is simply no maintaining any activity system without identity work. Alexander and Wiley (1981, p. 274) call identity formation the "cornerstone of interaction." We would call it the cornerstone of social life.

Secondly, identity work is necessary to meet the existential needs that social life engenders in individuals. Identities give us a sense of belonging, feelings of personal significance, a sense of location relative to others, a sense of continuity and coherence, and feelings of worth. It is through identity work that we create these distress-reducing experiences (cf. Hewitt 1989; Klapp 1969; Strauss 1959). By implication, ineffective identity work may lead to feelings of anxiety, isolation, insignificance, confusion, and inauthenticity. In our view, identity work is necessary to maintain emotional and cognitive equilibrium—or, more simply, sanity—because we must signify ourselves to others in a consistent enough way to generate consistent feedback about who and what we are. Identity work, in sum, is how we hold ourselves, each other, and the interaction order together.

Although we can, as individuals, artfully signify our unique subjectivities, even this identity work depends on our relationships with others. Collaboration is needed to create the signs, meanings, and codes that enable all kinds of identity work to go on. It is this essential collaborative activity that we are calling subcultural identity work. As we see it, this is the generic process through which people create and learn to use the symbolic tools needed to give facticity and coherence to themselves and others. We turn now to an outline of what the process of subcultural identity work entails.

The Process of Subcultural Identity Work

Based on our own studies and other literature on identity formation, we propose that there are four major parts to the process of subcultural identity work: (1)

defining, or the creation of a social representation that brings an identity into existence; (2) coding, or the creation of a set of rules for signifying an identity; (3)
affirming, or the creation of opportunities for enacting and validating claims to an
identity; and (4) policing, or the protection of the meaning of an identity and
enforcement of the code for signifying it. Although we imply a sequence here, in
real cases these parts of the process are likely to overlap considerably (as our
empirical illustrations will attest). Our claim is thus not that all cases of subcultural identity work unfold in this sequence but that *to be successful* the process
must include these parts.

Defining, as we see it, is a matter of creating a social representation (Farr and
Moscovici 1984), which here refers to a shared idea about the existence of a category or type of people. It is a matter of defining into existence not only a category
but also a concomitant identity that adheres to people who belong to the category.
Consider, for example, the category "white people" and the racial identity "white."
The category itself exists as a social representation or, as some might say, a shared
fantasy created for political reasons (Fields 1990; Omi and Winant 1994). This
social representation had to be talked or written into existence once upon a time.
Only after this work was done could anyone claim the identity "white" or "white
person."

The people to whom social representations apply might or might not be their
originators. Social representations are invented in many places—popular music,
advertising, academia, the news media, government, street life—and may "catch
on" from there, sweeping up people who had no idea they belonged to a category.
Consider, as examples, identities such as yuppie, gangsta, and post-modernist.
These are not labels for concrete groups but names given to widely scattered people who fit certain patterns of behavior or demographic slots. Yet once the label is
created, a social representation exists that can become the basis for a real and consequential identity. In other cases, people may first recognize their commonality
and then try to create a social representation for themselves to suit their own purposes.

Subcultural identity work thus begins with the creation of a symbol, image, or
name that represents, in more than one mind, a group of people. The representation may encompass people who share a physical trait, behavior, attitude, activity,
place, history, language, or social location (Tajfel 1981). Once the name exists and
takes hold as a social representation—which is to say, once the name is widely
used, with roughly shared understanding about who it refers to—both the group
and a new identity exist as parts of the culture. It then becomes possible for people
to signify themselves in new ways.

Defining includes the work of investing an identity with meaning. What does it
mean, to use a fictitious example, to be an *Elamite*? It could mean nothing more
than that you are from the newly created republic of Elam. Or it could mean you
are a member of the newly formed church of the prophet Ela; or that you are a
player of the newly invented game of El. These would merely be starting points,

however, for the investment and accretion of meaning. Further identity work, in other words, would be required to establish understandings about the qualities *implied* by the label *Elamite*. The Elamites might do this for themselves; others might do it to them.

Identities acquire meanings over time. In general, the process must involve linking signs of qualities with signs of the identity. This can be done through talk that calls attention to an act, declares the act a sign of certain qualities, and labels the actor. It might thus be said, for example, "This was a marvelous performance—a sign of immense talent and dedication. Such is typical of Juilliard students." Identities can also accrue meanings when the connections between act, identity, and character are implied, as, for example, when news images of African Americans consistently show them as street criminals or welfare mothers. Here, the connections between identity and character are based on tacit assumptions about the truth value and representativeness of news and the homogeneity of a group. The identity-defining subtext remains the same: "That's what those people are like."

Some identities are obviously more valuable than others in that they elicit more deference and respect. Groups that invent or remake high-status identities for themselves may have to do extra identity work to contend with the reactions of threatened groups. Defining, in such instances, may include strategic mythologizing.

What this means, to invoke the Elamites again, is that if they define *Elamite* as implying special virtue, ability, or worth, they will probably be challenged. Those who could be disadvantaged by the high status of Elamites might prefer to define *Elamite* as implying arrogance and a lust for power. To deflect such indictments, the Elamites might devise a myth in which Elamites are portrayed as deriving their noble qualities from nature or God (cf. Mol 1976). The Elamites could then claim to be what they see themselves as being—strong, rational, aggressive, suited for ruling—not just because they say so but because nature or God has made them so. Those who would dispute the glorious meaning of *Elamite* are thus exposed as fools or blasphemers, fit for exploitation or death.

History is replete with examples of just such mythologizing, or what is also called *essentializing*, which refers to the making of doctrinal claims that certain good or bad traits inhere in all who share an identity. For example, part of inventing the white race—as a way to forestall solidarity between Africans and working-class Europeans (Allen 1994; Frankenburg 1993; Roediger 1991; Saxton 1990)—involved fabricating beliefs in the biological basis of race and in innate "white" superiority. Men have done much the same kind of fabricating, promoting the notions that superior capacities for reason, organization, and control are among the natural endowments of males. Essentializing by appeal to nature (physiology or genes), rather than to religion, is the preferred modernist strategy for ideologically anchoring the meaning of politically expedient identities and policing their boundaries.

Once an identity is defined, a set of rules or enabling conventions for signifying the identity must be created. We will call these rules *identity codes*, the idea being that each identity has its own local code. To make a successful claim to an identity—to signify it properly—an individual must know, and know how to use, the code (cf. Wieder 1974). To put it another way, an identity code consists of practical knowledge about how to show that you belong to a group or category or have particular qualities. Identity codes also enable recognition and appreciation of others' identity enactments. Such codes are the human analog of the pheromones upon which animals and insects rely for identification of species mates.

Sometimes the definition of a category carries with it an implicit identity code. For example, if *baby boomers* are defined as people born in the United States between 1946 and 1964, then all one must normally do to claim the identity *baby boomer* is to claim a birthdate within that time frame. But even in this case, the claim to an appropriate birthdate must accord with other signifiers, such as appearances of appropriate age and citizenship. The need to signify these additional qualifications would also be part of the code implicit in the definition of the category.

Regardless of how simple or complex the rules for signifying an identity might be, the rules must be created and individuals must learn to use them. Boys learn, for example, that while having a penis is adequate to signify maleness, signifying *masculinity*—as a set of capacities and dispositions—requires facility with a more complex set of rules and conventions. Boys must learn how to dress, talk, and act so that they will be seen as possessing a self with masculine qualities. Girls and women must of course learn their own codes. In fact, in a pervasively gendered society, we must learn all the operative codes for at least *perceiving* the range of possible gender identities (Bem 1981).

As a kind of subcultural identity work, coding thus includes giving meanings to signs that individuals can then use to signify who and what they are. So, for example, a group must devise and transmit a set of understandings about the meaning of certain words, dialects, acts, pieces of clothing, hair styles, facial expressions, and anything else that can be "wielded" by an individual who wants to claim membership in the group. Individuals can take the initiative in trying to create such signs, but individuals alone can't make signs effective. This can only be done jointly, by creating a code that selects certain acts and objects as signs of a kind of self.

Like defining, coding can be contentious. When valued identities are at stake, people might try to bend the code toward giving greater significance to the signs they can most readily wield. Large males, for example, might be inclined to see physical strength as the key signifier of manhood; smaller males might prefer to attach greater significance to intelligence. Whatever the identity at stake, the code must say how to make an identity claim stick, and, by implication, how to test the claims of others. Because of the possibility of contention, identity codes may never be settled once and for all; they may be negotiated continually as people try to stretch and modify them and as outsiders try to co-opt them.

An identity code also includes rules about who has the right to adjudicate identity claims and bestow the identity in question. The code, in other words, specifies who it is important to impress, whose affirmation matters and how affirmation should be offered, who decides what is a passable enactment, and who has the right to police the boundaries of the identity. Identity codes may also specify meanings to be given to pretenders—those whose enactments fail because they don't know the code or can't muster the requisite signs. Among bicycle racers for example, cyclists who dress like racers, but are slow and inept, are known pejoratively as *freds*. The tacit identity code that is part of the sport's subculture tells how to distinguish real racers from freds and how to avoid signifying fredness. Equivalent rules of discernment and penalties for fakers would seem to be part of every group's identity code.[7]

Once an identity is defined and a code established, it is necessary to create opportunities for enacting and affirming it. People who wish to claim an identity must have chances to signify, to be seen signifying, to have their acts of signification interpreted as they wish, and to have their claims affirmed by an audience that matters. In some cases, affirmation may involve elaborate initiatory rites; in others, it might involve no more than a nod from the right person. However long, intense, or complex such interactions might be, the point is that people must somehow arrange to encounter each other in ways and places that allow for signification and affirmation to occur. Organizing and scripting such encounters is an important kind of subcultural identity work. Over time such work may be routinized, but still it must be done and done jointly.

Without affirming, the identity in question remains an ephemeral notion, connected to no real persons. Moreover, without opportunities—be they elaborate rites, formal meetings, or casual encounters—to signify and be affirmed, the identity cannot bring psychic and material rewards to individuals. It would remain a private fantasy. Affirming is thus a practical necessity. It is necessary for keeping an identity alive and functional, so to speak, both for individuals who want to claim it and for the groups that need it for cohesion.

It could be argued that policing, which we see as the fourth part of subcultural identity work, goes on throughout the process. Defining would seem to imply policing from the start, in that the creation of a social representation entails setting boundaries and saying who's in and who's out. Likewise with coding, the rules saying what kinds of enactments are acceptable and what kinds are not. Even affirming can include an element of policing, in that the wrong people can be kept out of encounters where they might otherwise have a chance to signify and be affirmed. Or, more simply, affirmation can be withheld; that, too, is a form of policing.

We agree that policing can go on at any point in the process of subcultural identity work. As suggested above, this would refer to any efforts to establish and maintain boundaries between those who are in—those who can rightfully claim an identity—and those who, by agreement, should be kept out. Since all identity work creates or maintains boundaries between individuals and groups, it could even be argued that identity work itself is synonymous with what we're calling policing. That argument is plausible though not very useful.

While it is true that all identity work could be construed as a form of policing, we think it's helpful to distinguish the work done to create an identity from the work done to preserve it as a piece of social reality. All identities require some such work to ensure their integrity and survival, since definitions are always subject to dispute and negotiation, codes can be forgotten or heretically altered, and rites of affirmation can be stolen, intruded upon, and disrupted. Intra- and inter-group struggles can also raise the demands for policing. Our claim is thus that policing, because of the inherently unstable nature of signs and codes and because of human conflict and creativity, becomes a project in its own right once an identity exists and people have a stake in maintaining it.

The seriousness of threats and challenges can, of course, vary tremendously, so that policing may require more or less time and energy. It may require only mild reassertion of definitions when meanings seem to drift. Likewise, codes can be firmed up by simple acknowledgments of exemplary identity enactments when routine enactments get sloppy. At the other extreme, in the face of more severe threats, definitions and codes may be reasserted with the force of crusades, inquisitions, and pogroms. Considering that group survival and satisfaction of the most compelling existential needs of individuals can be staked on identities, it is not surprising that humans will die or kill for them. *Policing* hardly seems like an adequate term in such cases.

Although the term *policing* suggests that it is always a high-status group doing subcultural identity work to protect the value of its symbolic resources, any group can undertake policing to protect a shared identity. Policing may be especially vigorous in subordinate groups if the identity at stake is one that aids group survival in the face of a hostile society. By our definition, however, policing could also include resisting stigma and breaking oppressive identity codes. In recent decades many previously stigmatized groups have tried to do exactly this—that is, to take control of the meaning of their identities away from the dominant group. Later, we will discuss this in terms of oppositional identity work.

Although individuals can do policing on behalf of others with whom they share an identity, policing requires collaboration to be effective since it is consensus that holds the line against a breakdown of the code or loss of an identity's meaning. Policing is thus likely to entail debate, politicking, and organizing—all of which are collaborative activities that, if successful, also serve to preserve the group from which a shared identity derives. Again, individuals can take policing *initiatives* by calling attention to the violations of others (to do so is often a form of individual identity work because to notice a violation implies that one is a genuine insider, hip to the proper code). But the *force* of policing stems from joint action to remake a frayed or besmirched identity.

Our sketch of the process of subcultural identity work suggests how the study of identity work can move beyond concern for the situated self-presentations of individuals. We have proposed a closer study of how identities are collaboratively created as social representations, how they are invested with and accrue meaning over time, how identity codes are created and transmitted, how opportunities for signifying and affirming are created, and how people who share an identity collaborate to protect it. So far, however, we have offered only a schematic theoretical account. In the next section, we offer two empirical illustrations of subcultural identity work.

EMPIRICAL ILLUSTRATIONS

The perspective on identity work set forth above can highlight certain otherwise invisible aspects of group process. To show how this is so, we discuss two examples of subcultural identity work, based on our own recent studies of groups formed, in part, to cope with identity-related problems (Mason-Schrock 1995; Schwalbe 1995, 1996). Because of the peculiar nature of the problems dealt with by these groups, not every aspect of subcultural identity-making was of equal concern to them. Our case studies thus do not illustrate the whole process in detail but rather offer a sample of the kind of analysis that can grow out of a focus on subcultural identity-making.

Both groups—a mythopoetic men's group and a support group for transsexuals—were dealing with parallel kinds of identity problems. In both cases the problems had to do with the moral implications of gender-related identities. The mythopoetic men were trying to invest the identity *man* with positive moral value, that is, they were trying to define their shared identity *man* as implying special virtues. The transsexuals were struggling to make the identity *transsexual* a morally acceptable one. In both cases the identity-making process involved some defining, coding, affirming, and policing.

The Mythopoetic Men

These men were admirers of the poet Robert Bly, author of the 1991 bestseller *Iron John*. They were predominantly white, college-educated, middle class, and middle aged. Bly's teachings, drawing heavily on Jungian psychology, urged the men to break out of their mid-life ruts and reinvigorate themselves by reclaiming their "masculine energies," especially those associated with the "wildman" and "warrior" archetypes. The therapeutic regimen through which the men tried to do this included drumming, dancing, mask making, ritual, myth, and poetry—usually undertaken by all-male groups in rustic settings. The activity was called *mythopoetic* in reference to its elements of myth and poetry. Although this activity, carried on in workshops and retreats across the country (beginning in 1981), never constituted an organized social movement, that's how the media tagged it in the late 1980s and early 1990s: the men's movement.

In the fall of 1990, the first author began a study of a group of local men engaged in mythopoetic activity. Data came from participant-observation at 128 meetings of various kinds over a three-year period, interviews with 21 of the core group of approximately 30 men most heavily involved in the activity, and analysis of documents produced by movement leaders and by other mythopoetic men's groups around the country. The study is described in detail elsewhere (Schwalbe 1996).

One thing that men drawn to mythopoetic activity were seeking was relief from guilt induced by the meanings that the identity *man* had accrued as a result of feminist criticism in the 1960s, 1970s, and 1980s. For various reasons, these men were sensitive to claims by some feminists that men are naturally unfeeling, violent, incapable of nurturing, and thus morally inferior to women. The men perceived this criticism as an unfair indictment of the moral worth of all men. It seemed especially unfair since it made no exceptions for gentle men like themselves, many of whom had long supported equal rights for women. Part of what the men were thus seeking was assurance that, as men, they were not bad people. This called for redefining the meaning of *man* as an identity; more specifically, it called for remaking *man* as a moral identity.[8]

Analysis of mythopoetic activity suggests that it was, in large part, an identity reconstruction project. Remaking *man* as a moral identity was not incidental but central to what mythopoetic activity was about. In terms of the argument here, mythopoetic activity was in large part subcultural identity work. What follows is an abbreviated account of some of the ways in which the men collaborated to do this work at their retreats and other gatherings.

Defining Through Forms of Talk

By using certain phrases the men reaffirmed and revalued their shared identity as men. For example, the men called their activity "men's work," thus implying not only that they were indeed men, but men engaged in a serious effort to get something done. The men also reminded themselves of their shared identity through the frequent use of the tag phrase "as men." For example, in talking about myths and fairy tales, it might be said, "These stories have a great deal to teach us about our lives as men." Or, if the topic were respect for the earth, it might be said, "We need to think about what it means to care for the earth, as men." This self-reminding about being men was important, too, because the men were engaged in activities—sharing feelings, same-sex dancing, affectionate nurturing, sometimes weeping—that, by usual cultural standards, would have seemed quite "womanly."

At their gatherings the men avoided contentious topics. In fact, any statement that suggested an opinion was usually prefaced with the disclaimer "for me," as in "For me, the Gulf War was very depressing." This disclaimer signified that a man was speaking for himself and wasn't insisting that other men had to agree. It was revealing, however, that the men never used this disclaimer when declaiming about the goodness of men. It might thus be said, "One thing I've learned from doing men's work is that there's a lot of good in men," or "By being in this group, I've seen that men are trustworthy and capable of nurturing." Such statements were never qualified with "for me" since to use this phrase would have implied that an opposing point of view was reasonable. In this way—by not qualifying positive statements about men—the men subtly raised the moral value of *man* as a shared identity.

Part of what the men were trying to do was to redefine *man* as implying the qualities of emotional depth and sensitivity, compassion, an ability to nurture, a desire to husband the resources of the earth, strength of conviction, and generativity. It was important to the mythopoetic men to resist the moral indictment of feminism and to redefine *man* as a sign of a male self that was virtuously assertive rather than violent and brutish. The forms of talk noted above were one way this redefining was done. But the men also did it through other group practices, such as the telling of poems and fairy tales.

Defining Through Group Practices

Most gatherings included some reciting of poems. Usually these had to do with father-son relationships, nature, spirituality, or, less often, relationships with women. On one level, the poems were used to elicit feelings and images as part of the quasi-Jungian therapy that was going on. On another level, they served to demonstrate men's capacities to appreciate and be moved by beauty. At gatherings, men often responded to poems with heavy sighs and deep mmmmms, or loud applause, whoops of assent, and calls for a poem to be read again. It was thus signified that men were not emotionally inert nor aroused only by power. Leaders and participants often remarked on these reactions as proof of men's complex aesthetic and emotional sensitivities. Such comments ensured that no one missed the new layers of meaning that were being added to *man*.

Myths and fairy tales were also told at gatherings. These were usually drawn from the ancient Greeks, African and Native American folk traditions, the Grimms' collection, or the Arthurian legends. Most of the stories featured a male hero with whom the men hearing the story were invited to identify. In mythopoetic lore, the hero was understood to be a figure representing masculine archetypal energies. A strong, wise, resolute, and compassionate hero was thus not simply an appealing role model but a symbol of the powers inherent in all men. By implication, the qualities displayed by the hero were qualities potentially accessible to all who shared the identity *man*. The identity *man* was thus defined as signifying a self with enormous potential for noble action.

After a story or a segment of a story was told, the men talked about the images it contained, especially those that evoked strong emotional responses. Again, this manner of talking about the stories was part of the Jungian strategy of using emotionally evocative images to access the unconscious. During these discussions it usually turned out that, while many different images were mentioned, many men cited the same images as having resonated with them. The manner in which this pattern of responses was interpreted helped to accomplish a bit of identity work.

The *overlap* in reactions was seen as evidence that men indeed shared a psychic connection with all men, thus implying that Jungian ideas about archetypal masculine energies were true. The *diversity* of reactions was seen as evidence of the emotional complexity of men, thus implying, as did the men's responses to poems, that men were very much alive and unique inside—not the uniform bags of emotional sawdust suggested by some feminist theory (as the men perceived it). What's important to see is that this method of discussing the stories created further openings for adding positive moral value to the identity *man*.

Coding and Policing

Jungian psychology was a key ideological resource for the men and, especially, for the movement's leaders. Jungian psychology held that women and men were essentially different in that archetypal masculine energies (i.e., psychic energies) flowed more powerfully through men than women, and that archetypal feminine energies flowed more powerfully through women than men. Social experience was said to be necessary to fully activate the archetypes that gave form to psychic energies, but the archetypes themselves were innate. By implication, if a man didn't show all the qualities of mature manhood, this was not because his character was defective but because his natural masculine powers had yet to be brought "on line." The mythopoetic perspective thus taught that the best qualities of manhood inhered in the selves of all men.

The mythopoetic men thus drew on Jungian psychology to anchor the meaning of *man* in a transcendent realm of archetypes. According to Jungian psychology, men were what they were because evolution had endowed them all with certain archetypes. The selves of men thus could not be discredited by feminist criticism, no matter how badly some men behaved. Jungian psychology also drew boundaries based on sex. Supposedly, males and females were archetypally endowed differently by virtue of being male or female, not by virtue of socialization. Boundaries could thus be policed by applying simple tests. But in this case the point wasn't so much to keep women out as it was to include all males. Having a penis was adequate qualification for claiming the identity *man* and its concomitant virtues.

Jungian psychology aided the policing function in another way: it held that the archetypes of mature masculinity could be activated in men only by other men (usually in initiatory rites). This meant that only men could do the work of making other men—building, of course, on a biological foundation. The implication was that while women, in corollary fashion, could activate feminine archetypes in other women, they had little to say about the making of men. It was simply not within women's powers or purview, according to Jungian psychology, to do the job. As a matter of gender politics, this established men as the community with the exclusive right to define the meaning of *man* and to decide who could rightfully claim this identity.

Affirming

The identity code the men created went beyond Jungian psychology. In the mythopoetic subculture it was important to signify not only manhood by virtue of maleness but also *mature manhood*—or what was also called "deep masculinity"—through behavior. Such signifying behavior included freely showing affection for other men in the forms of hugging and touching, speaking one's personal "gut truth" and sharing one's feelings, and engaging in uninhibited dancing and situationally appropriate silliness. When men did these things at gatherings it was partly a matter of individual identity work—a displaying of one's progress on a path of personal growth. But such behavior also helped to reinforce an identity code that was an integral part of the group's subculture.

Mythopoetic philosophy held that while men could indeed pursue their spiritual quests as individuals, they needed the company of other men for support, affirmation, and opportunities to engage in rituals that would help bring their masculine energies on line. Again, this thinking derived in large part from Jungian psychology. The practical implication was that the men had to gather by themselves to help each other meet their supposedly unique psychic needs as men. What this provided was opportunities for affirmation. Gatherings, in other words, were opportunities for men to encounter each other, to define and redefine the meaning of *man*, and to affirm that all those present had a right to claim this identity.

Jungian psychology thus provided an ideological resource for defining the identity *man*, for establishing an identity code, for establishing the importance of gatherings, and for policing the boundaries of the identity that the men were trying to remake. It was the mythopoetic leaders who took the lead in using Jungian psychology in this way (often without many of the participating men being aware of it). Ideas from other humanistic psychotherapies, New Age mysticism, and pantheistic religions were also added to the Jungian core. Moreover, the men did not just borrow ideas from other places but invented some of their own and also devised ways to put these ideas into practice at gatherings. What they assembled was a hodgepodge of comforting ideas about gender, the psyche, the meaning of *man*, and about how to signify manhood.

The Transsexuals

Part of becoming a transsexual was coming to believe that the gender of one's "true self" was different from the gender signified by one's body. This belief was not easy to arrive at, since it clashed with the common presumption that the sex of the body is an unequivocal sign of gender. Even people who feel intensely inauthentic when doing gender conventionally do not usually see themselves as men trapped in female bodies, or women trapped in male bodies. The idea that such a condition is possible has been around for a long time, however, and is embraced by members of what has been dubbed the "transgender community," which consists of people who, for various reasons, enjoy wearing gender-discrepant clothes (i.e., "crossdressing") and people who are, or who are coming to define themselves as, transsexuals. The transgender community is diffuse, being constituted largely by over 200 local support groups in cities around the United States.

Beginning in the fall of 1992, the second author studied one such group for 15 months. During this time, eight three-hour support group meetings and a Christmas party for members of the group and their partners were observed. The researcher also attended a hockey game with a born male dressed as a woman. Nine male-to-female transsexuals and one female-to-male transsexual, all members of the support group, were interviewed at length. Further data came from transgender community publications and from postings in Internet discussion groups for transsexuals and crossdressers. A detailed account of the study can be found elsewhere (Mason-Schrock 1995).

Defining

In interviews, support group meetings, and online forums, transsexuals told of having long felt inauthentic in enacting their prescribed gender identities. They also frequently talked about questing fruitlessly for a "true self." Many said that they lived this way for years, not knowing what was wrong with them, trying various remedies. It was only when they began to suspect that they might be transsexual that their experience began to make sense. When they finally found their way into it, the transgender community provided them with an interpretive scheme for explaining their experience, opportunities to interact with others whose experiences were similar, and a program for changing identities. It was a subculture that had evolved to accomplish the unusual kind of identity work that transsexuals found it hard or impossible to do alone.

Transsexuals were seeking not just authenticity and a radical identity change. They were also trying to normalize the identity *transsexual*, that is, they wanted to make it morally acceptable. This was difficult because, by violating one of the basic precepts of the culture's gender order—that gender should correspond to biological sex—transsexuals were stigmatized as sexually perverse or mentally ill. Their subcultural identity work thus also had to deflect these moral indictments and allow them to feel that they were decent people, despite their violation of the gender order.

An important ideological tool for doing this identity work was the widespread notion that gender is biologically determined. Transsexuals found this idea to be a useful stigma shield because it implied that transsexuality was not a perverse individual choice but a matter of biological miswiring. Their belief was that, while nature normally caused males to become men and females to become women, nature could also make mistakes. Transsexuality was thus no more legitimate a basis for stigma than, say, a harelip. Although this belief in the biological natural-ness of gender was readily available in the culture at large, transsexuals nonethe-less invested extra efforts in reinforcing it among themselves. This was part of defining what it meant to be a transsexual.

Transsexuals reinforced this biological view of gender through the careful and consistent use of small bits of language. For example, they used phrases such as "when I dress as a female" and "my female self has always existed." When they discussed the synthetic hormones that many of them took, these drugs were referred to as "female hormones." The transition from living as men to living as women was described as "puberty." Recruits to the transgender community learned that this was how one should speak of these matters. Such language func-tioned to naturalize gender and, by implication, to destigmatize the identity of those who felt compelled to correct nature's mistake.

Coding

Although naturalizing gender helped transsexuals preserve a sense of moral worth, it created another problem. If nature was responsible for implanting gen-dered selves at birth, then transsexuals needed proof that their "true selves" had always been mismatched with their bodies. Transsexuals needed proof, in other words, that they had always been differently-gendered inside. The transgender community taught that such proof could be found in a person's biography, if one knew what to look for. To get at this kind of evidence, support group members were encouraged to tell stories about their earliest experiences of gender unease. This kind of story telling and the collective reinterpretation of biography were sta-ple parts of support group meetings.

Recalling early childhood events was a key part of this story telling practice because transsexuals believed that at this time in one's life the "true self" had not yet been stifled by the gender norms of society. The stories most often told were about crossdressing. Having worn gender-discrepant clothes, whether once or many times, or merely having fantasized about doing so was construed as a sign of a differently-gendered true self. Born males also often told stories about failing at sports as children. Born females talked about climbing trees, getting in fights, and other forms of tomboyish behavior. Support group members encouraged each other to see these stories as evidence of having been born in the wrong-sexed body.

Stories about being "in denial" later in life were also important for coming to terms with the identity *transsexual*. These kinds of stories explained away experi-ences that might have suggested a different kind of true self. For example, born males often went through stages where they engaged in either hyper-masculine activities (heavy sports involvement, gun collecting, motorcycling) or thought of themselves as gay or as crossdressers. Accounts of such experiences were typically preceded or followed with the disclaimer, "when I was in denial." The implication was that the true self, discovered only later, was present at an earlier time, despite a person's efforts to deny it. Stories of having been in denial not only signified a transsexual identity, they also provided a sense of personal continuity in the face of a radical identity change.

Support group members encouraged each other to tell stories about childhood gender violations and about denial. The encouragement often came in the form of leading questions. For instance, new group members (including the researcher) were usually asked about crossdressing in childhood. On two occasions when several newcomers were present, the facilitator asked for a show of hands from people who had "purged their female clothing." Both times, most people raised their hands, and both times an experienced group member said that when *she* had been in denial she had purged repeatedly. The lesson for new members was that talk about purging was not frivolous but, when linked to the notion of denial, a way to signify the identity *transsexual*.

New members also picked up the group's line of talk by modeling others. Often a member would say something like, "I've felt like I was different from other boys as far back as I can remember," and then proceed to talk about early crossdressing or failing at sports. Such accounts showed new members that telling stories about gender nonconformity could be used to claim a transsexual identity. But it was the reactions of others that cemented this. If a male transsexual said s/he was always the last one picked for team sports, other experienced group members would nod, smile, and jump in with similar stories about themselves. Modeling and group encouragement thus taught new members how to signify transsexuality. This was, in other words, how the identity code was taught.

Affirming

In addition to defining what it meant to be a transsexual, and developing a repertoire of ways to signify a transsexual identity, transsexuals also had to make the identity morally acceptable. Naturalizing gender, as discussed above, was a key part of this. But transsexuals also redefined the moral meaning of their shared identity by equating themselves with other oppressed minorities. At support group meetings, transsexuals sometimes talked about how they faced discrimination akin to that experienced by gays, lesbians, and African Americans. One time a transgendered lawyer spoke to the group about the "civil rights of transsexuals," thus affirming the link between transsexuality and minority-group status. Through this positioning of themselves as a legitimate minority group, transsexuals both asserted their own goodness and implicitly cast their detractors as bigots.

The group thus not only helped transsexuals arrive at a belief in a differently-gendered true self; it provided them with means of expressing it, a moral schema for revaluing it, and opportunities to have it affirmed. In addition to the support group meetings, most participants were active in Internet newsgroups and America Online forums—places where they could also signify and be affirmed. Regional and national conventions provided still more opportunities. For many transsexuals this was the main function of the transgender community: a chance to interact with others who would affirm and valorize the identity *transsexual*.

Policing

Transsexuals protected the moral value of their shared identity by avoiding talk about sexuality. This was important to offset the stereotype of the person who crossdresses as a pervert. Sexuality came up at only one support group meeting—when a transsexual said she had to explain to a friend, to whom she was coming out, the difference between homosexuality and transsexuality. Sexuality was also avoided by screening prospective group members. When first approached about the research project, the facilitator explained that the group was for those who crossdressed for "purer reasons" than sexually motivated transvestites. This can be seen as a form of policing. By excluding those who crossdressed for erotic reasons, and by discouraging talk about sexuality, transsexuals deflected a damning stereotype.

Policing also occurred as transsexuals distinguished themselves from crossdressers. This was important because most transsexuals had for a time labeled themselves crossdressers. Becoming a transsexual meant distancing one's self from this previous identity. The boundary between the two identities was policed in several ways. For example, when members introduced themselves at meetings, they were expected to claim one identity or the other ("I'm Taylor, and I'm a crossdresser," or "I'm Leslie, and I'm a transsexual"). If newcomers were confused about which identity to claim, they would be counseled by the group and encouraged to make a decision. Boundaries were also reinforced through discussions of motivations for crossdressing. People who said that crossdressing allowed them to express the feminine *side* of themselves were seen as crossdressers. Transsexuals, in contrast, emphasized that, for them, crossdressing was a way to express a true self, one that was essentially and totally of a different gender than their bodies signified.

What transsexuals learned through their involvement in the transgender community was what it meant to be a transsexual, how to scan one's biography for evidence of transsexuality, how to interpret that evidence, how to present themselves as transsexual, how to affirm others' claims to this identity, and how to distinguish transsexuals from transvestites and crossdressers. In terms of our previous discussion, what they participated in, as members of this subcultural community, was the process of defining the identity *transsexual*, signifying it according to the transgender community's identity code, creating opportunities to have this identity affirmed, and policing its boundaries.

Other Studies

Two other recent studies of identity work in subcultural groups provide helpful illustrations of how the process occurs. Thumma (1991) studied an organization called Good News, which was formed to help gay Christians reconcile their sexuality with their religion. Thumma's analysis shows how the group helped people to reinterpret scripture in a way that allowed them to feel good about themselves as Christians, despite a sexuality that most Christians saw as sinful. The group's strategy was to define "being gay" as a result of God's will, while defining moral condemnation (from fundamentalists) as "men's fears and opinions being spoken in God's name." The subtext here, for those trying to reconcile Christianity and gayness, was, "God made you what you are and accepts you that way; knowing that, you can now get on with developing your spirituality."

We see in Thumma's study echoes of themes found in our studies of the mythopoetic men and the transsexuals. In each instance, the subgroup provides an interpretive scheme that allows people to have it both ways: they can be feminine *and* deeply masculine; they can believe in gender as a natural fact *and* yet proceed to socially reconstruct it; they can be Christian *and* gay. The general principle seems to be that identity dilemmas are what give rise to subcultural identity work. When the dilemmas are too much for individuals to handle, they may turn to others for help. It is then through joint action that people create and sustain the interpretive schemes, facilitating discourses, and rites of affirmation that allow them to live with, if not fully reconcile, the contradictions between identities they cannot or will not abandon.

In his analysis of identity construction among gay men with AIDS, Sandstrom (1990) shows how the men collaborated to reject both stigma and the imputation of victimhood. Through their joint action, the men redefined the identity *person with AIDS* as implying special insight, purpose, and strength. Sandstrom shows how subcultural involvement provided "the social and symbolic resources necessary to fashion revitalizing identities and to sustain a sense of dignity and self-worth" (p. 293). It's important to note that in this case the men's identity work was not purely ideological or a matter of group talk; they affirmed the qualities implied by their remade identities by dedicating themselves to activism.

Sandstrom's study suggests that any identity can be made into a moral identity. As the mythopoetic men sought to do with *man*, which they felt had been besmirched, so the gay men sought to do with *person with AIDS*, an indisputably stigmatized identity in the culture at large. Sandstrom's analysis also shows how this kind of identity work depends on joint action to create symbolic resources that may not yet exist, to teach new members how to use these resources, and to give support in the face of a hostile culture. We also see here identity work being done through highly visible redemptive acts. It was important that the men saw each other doing good work (e.g., community education). This was powerful affirmation of the moral character implied by *person with AIDS*. Sandstrom's study also shows how subcultural identity work can remake not only meanings, through talk, but lives, through action.

IDENTITY-MAKING AS CULTURAL STRUGGLE

The focus on subcultural identity work moves us from the individual to the meso level. Instead of asking how individuals use what is in their identity kits, we have asked how those kits are created and maintained. To call this a *sub*cultural process implies, however, that the kit making is shaped by the world around it. So far we have only hinted at the nature of such subgroup/society connections. In this final section, then, we want to suggest how social psychology can reclaim its grip on identity—as a sociological phenomenon—by examining identity-making in cultural and structural context.

Structural interactionism links identity to social structure by way of the notion of role (McCall and Simmons 1978; Stryker 1980). In this view, identities—seen as meanings given to the self—derive from role performance. Self and society are linked because roles are seen as part of the organizing scheme of society. A person's constellation of identities thus mirrors, at least in part, the pattern of roles present in society. To us this seems a rather static view. We prefer to emphasize the ways in which material conditions and culture affect identity work. This makes it possible to see how individual and subcultural identity work are enabled and constrained by the conditions under which people act.

It has been implicit in our argument that all social life requires identity work. But we would also argue that the work may be more or less arduous and constant depending on features of the context. In a society where it is possible, perhaps inevitable, to be many different things at once, people may have to invest considerable effort in signifying who and what they are in an encounter. Status inequalities may also amplify this need, as serious consequences can ensue from establishing or failing to establish one's claims to a valued identity. Intergroup conflict might likewise call for intense identity work. Under such conditions, a great deal of subcultural identity work is likely to occur as groups lay claim to valued identities, or revalue ones they already possess, while perhaps also trying to impose devalued identities on their antagonists.

The conflict, complexity, inequality, and cultural diversity of many modern societies (or postmodern, as some would have it) seem to generate demands for almost constant identity work. When the chances of being mislabeled, slighted, or dominated are great, it is crucial to be able to effectively claim the most powerful and valued identity one can. When dominant ideologies equate individual worth with wealth and power, impression management may be necessary just to establish visibility and presence as a person. When market forces homogenize culture, when hierarchical institutions squelch creativity, and when political arrangements prove resistant to protest, many people may struggle to create identities that sustain feelings of personal significance, specialness, and agency. A focus on identity work can help us understand how these conditions shape what it is that people actually do together by way of making identities, rather than what they do by way of privately claiming labels for themselves.

Much so-called postmodern social theory has grown up in an attempt to explain how the conditions of contemporary social life impinge on individuals. Theorists writing in this genre have made liberal use of the concepts of self and identity (see, e.g., Calhoun 1994; Lash and Friedman 1992; Taylor 1989). But generally, they have done so without much awareness of the research and writing about self and identity in social psychology.[9] As Gecas and Burke (1995) argue, social psychologists interested in links between self, identity, and society can find stimulating insights in the work of postmodern social theorists. Taking those insights into account, we should also continue to advance our own analyses of these matters, building on well-developed traditions of theorizing about self and identity.

Our emphasis on identity work as group process is an attempt to advance our understanding of how identities are jointly made, of how and why individuals use them as they do, and of how it is that identities, once they are created, act back on social life. To focus on identity work is thus to examine the place where people, through expressive behavior and face-to-face interaction, both reproduce and resist larger social arrangements. So far, we've done this by emphasizing action at the level of the cultural subgroup. We want to conclude by offering several concepts that suggest how the analysis of identity work can be taken a step farther, so as to give us a grip not only on identity but on the context of its making.

Oppressive Identity Work

In any encounter, people may claim identities for themselves and impose identities on others. The latter has been called altercasting (Weinstein and Deutschberger 1963). Part of subcultural identity work may be altercasting whole groups of others. Hunt and Benford (1994) describe this in their study of the peace and justice movement. Activists sought not only to create identities for themselves but for other actors in the same political arena. All identity claims logically imply difference from others; this in turn implies a characterization of others as lacking some quality or qualities possessed by the original identity claimant. Generically speaking, we can call this process "othering" (Fine 1994, pp.78-79). Under conditions of inequality, however, othering is likely to be pernicious.

To use an example used earlier, racial identities are not politically innocent matters of difference. Historically, they tend to be created as social control devices by dominant groups (Deschamps 1982). Racial categorization schemes thus do not simply say, "There are different kinds of people," but rather, "There are different kinds of people, all of whom are inferior to us in some way." When a dominant group engages in this kind of othering—a casting of members of other groups as less worthy and capable human beings, if human at all—it constitutes what we call oppressive identity work. As in the case of "race," the groupings themselves may be created as a matter of political expedience. In other words, oppressive identity work may define into existence a group of exploitable others.

Our foregoing analysis of subcultural identity work suggested that essentializing is a strategy often used to cement the otherwise purely symbolic reality of group differences. This can aid oppressive identity work by making the failings of others seem rooted in biology or divine law. But whatever the origin of the allegedly flawed selves that supposedly inhere in members of a devalued group, their status as members in such a group must be signified somehow. Oppressive identity work thus also requires creating an identity code that makes others identifiable as flawed. This involves the designation of what we call *peremptory signifiers*, the best example of which, at least in U.S. society, is skin color.

Peremptory signifiers—and these are, from the dominant group's perspective, signifiers of essential character—are imposed on others. Oppressive identity work is thus oppressive not only because it devalues others but because it denies them the power to signify their own character. Peremptory signifiers, in other words, cannot be sloughed off; they are ideologically fastened to subordinated others. This means that all expressive behavior by members of a subordinated group is interpreted in light of the peremptory signifier, which shapes the meaning of all other signs of character (cf. Goffman 1963, pp. 3-6). In this way the selves of subordinated others are profaned even before they can be volitionally signified.

Earlier we said that identities are indexes of the self. By this we meant that identities are complex signs of the capacities, worth, inclinations, and feelings that supposedly inhere in a person. To claim an identity is thus to claim a particular kind of self. As we have shown, however, the way in which any identity is "read" as an index of the self (i.e., an index of essential character) is a matter of negotiation, perhaps struggle. In creating the codes that say how identity claims and other types of expressive behavior should be read, a group creates possibilities for people to express particular kinds of selves. What a dominant group can do, by creating a code that limits the expressive agency of subordinated others, is to make it impossible for them to create fully creditable selves.

All this suggests a need to include in any analysis of subcultural identity work a consideration of matters of power and ideology. To understand how groups are able to create and use certain kinds of identity-making resources and not others, we must consider how those groups are positioned in the power structure of society. With regard to oppressive identity work, it is especially important to consider how some groups may be able to use the state to impose identities on others. To understand how and why individual members of these groups are or are not able to take control of their own identity work—in encounters on a daily basis—likewise requires considering the larger ideological context in which such action occurs. How else to understand the meaning of dark skin, a female body, or cheap clothes?

Oppositional Identity Work

To resist the stigma imposed by a dominant group, members of subordinated groups must engage in oppositional identity work. This is a matter of trying to transform discrediting identities into crediting ones, that is, to redefine those identities so they come to be seen as indexes of noble rather than flawed character. Although they were hardly oppressed, the mythopoetic men were doing something like this when they sought to remake *man* as a moral identity. The transsexuals likewise sought to destigmatize their shared identity by denying that *transsexual* implied a sexually perverse character. Individuals can, in various ways, resist stigmatization in their daily lives; but effective oppositional identity work must be a subcultural project, so as to create the resources, codes, and rites of affirmation that individuals need to signify better selves than the dominant culture permits.

Oppositional identity work often involves the subversion of a dominant group's identity codes. It can also be a process through which the creative energy of a group is focused and the artistic talent of individuals developed. Through oppositional identity work, subordinated groups not only resist their devaluing at the hands of the dominant group, they create themselves as people, individually and collectively. It is also important to see, however, that the compulsion to create a sense of worth through oppositional identity work can have unintended negative consequences. Two examples will help illustrate this.

Richard Majors (Majors and Billson 1992) has written about the "cool pose" of young black men in the inner city. This pose involves an ensemble of dress, talk, movement, posture, and facial expression intended to signify a self that is indomitable, capable of violence, and unfazed by much at all. Cool posing is thus a kind of identity work that individual men learn to do. But the possibility of doing it at all, let alone doing it well, requires jointly creating and then drawing on an identity code. Cool pose is also a message to white men, a message that Majors translates as, "White man, [the street] is my turf. You can't match me here." In this sense, it is a form of oppositional identity work. It is intended to signify a masculine self that is not passive and disempowered despite the lack of other signifiers, such as a prestigious job, high income, or college degree.

Cool posing is clearly a defensive response to a hostile society and can be appreciated as an artful assertion of agency. But as Majors and others (Anderson 1992; Staples 1978) have pointed out, cool posing can also lead to deadly character contests, emotional distance from loved ones, the abuse of women, and the reproduction of stereotypes that sustain racism. What is oppositional and revaluing in one way thus becomes destructive in others. It is a hard trap the men are in. The gender order of society requires males to signify a masculine self through various expressions of power and control. Yet, young black men are denied access to legitimate means of signifying a creditable masculine self. They also correctly perceive that to give up a claim to manhood would cost them one of the few sources of worth they have. Under these conditions, it is not surprising that there has arisen an identity code fraught with contradictions and the potential for bad results. Willis (1977) and MacLeod (1987) have shown that lower- and working-class white men are caught in much the same identity bind (see also Sennett and Cobb 1972; Tolson 1977).

A second example comes from research by Ogbu and Fordham (Fordham and Ogbu 1986; Ogbu 1978), who have looked at how black children try to define the identity *black* in terms opposed to the dominant white culture. They describe the situation:

> As a protection against the persistent assaults to self-esteem that are inherent in ghetto life, black street culture has evolved to legitimate certain behaviors prevalent in the black community that would otherwise be held in contempt by white society. Black identity is thus constructed as a series of oppositions to conventional middle-class "white" attitudes and behavior. (1986, p. 181)

As Fordham and Ogbu go on to explain, being *black* thus requires doing the opposite of what whites are perceived as doing, such as speaking standard American English, succeeding in school, working hard at routine jobs, marrying, and supporting their children. Black children who do these things may be ostracized for "acting white" (an example of policing).

Again it is possible to see the potential for harm. Solidarity and self-respect are maintained at the cost, in some cases, of educational success that could lead to better jobs. But as with cool posing, there is a trap here as well. To reject the oppositional definition of *black* and to try to succeed on the terms of white society is to risk rejection by both white society *and* one's community of origin. That is an enormous risk considering the impossibility of escaping skin color as a peremptory signifier. In this situation, then, oppositional identity work makes racial oppression bearable but may have the unintended consequence of reinforcing racist stereotypes, limiting individual achievement, and undermining more effective forms of resistance.

The generic problem described here does not arise solely with regard to race. The problem is one of redefining an identity so that it can serve as an index of creditable character, even when powerful others would say otherwise. Racial and ethnic minorities, women, gays and lesbians, the differently-abled, and working-class people have struggled to do precisely this kind of oppositional identity work. Such work can succeed in the sense of providing individuals with feelings of worth, a code for signifying a creditable self within a community, and opportunities for affirmation. It can also succeed in the sense of creating a more just society by dispelling images that diminish the worth of whole groups of people.

CONCLUSION

The bulk of social-psychological writing on identity treats identities as elements of the self-concept. In contrast, we have treated identities as semiotic constructs—indexes of essential character—that are created in interaction. We have thus emphasized processes of identity work rather than identities as reified pieces of culture or as meanings that individuals give to themselves. Most other analyses of identity work have focused on the expressive behavior of individuals. Our focus has been on the process of subcultural identity work, that is, the process whereby groups create the resources that make individual identity work possible. In the immediately preceding section, we tried to take the analysis a step farther to show how subcultural identity work can be better understood by taking culture and social structure into account.

One proposition at which we have arrived is that oppressive and oppositional identity work are the core processes of cultural struggle. Out of this struggle emerge the most consequential meanings created in any community. These are the meanings given to people, meanings that determine the value of the selves imputed to them. Through countervailing forms of identity work, people create themselves as human and, unfortunately, often create others as less than human and thus fit for exploitation, abuse, or segregation. To understand how and why such relations arise and evolve, we must regain our grip on identity and more deeply explore the intra- and intergroup processes through which identities are made.

ACKNOWLEDGMENTS

The authors thank Sherryl Kleinman, Viktor Gecas, Clifford Staples, Andrew Weigert, Joane Nagel, and the series editors for helpful comments on an earlier version of this chapter

NOTES

1. Another view sees the establishment and maintenance of situated identities as less problematic. For example, as McCall and Simmons (1978) see it, "The establishment of identities within an encounter is usually little more than a necessary prerequisite to the execution of other social tasks" (p. 143). In our view, shared by others cited in the text, the imputed character of the self is never safely fixed once and for all; one discrepant sign can dent or destroy it. Theorists who suppose that identity work is less pervasive than we do seem to forget how much self-signification and self-monitoring we do habitually, beneath conscious awareness, even in routine encounters.

2. Semiotics goes beyond symbolic interaction and dramaturgy to explore more deeply the nature of signs, how they work, and how they are used. For discussions of how semiotics can enhance interactionist analyses, see MacCannell (1976), Rochberg-Halton (1982), Manning (1987), and Manning and Cullum-Swan (1994).

144 MICHAEL L. SCHWALBE and DOUGLAS MASON-SCHROCK

3. In some cases, people's identity-making resources may slip away, requiring special forms of adaptive identity work. Kathy Charmaz (1987) has studied this process among the chronically ill.

4. The connection between language and the self is a long-running concern in pragmatist and interactionist thought. For analyses of this connection from an interactionist perspective, see Schwalbe (1983) and Denzin (1988). For analyses that take a semiotic tack, see Singer (1980) and Sebeok (1979).

5. Melucci (1989) also uses the term *collective identity*, but his definition is even farther from ours. He says, "Considered as a process, collective identity involves at least three fundamental dimensions which are in reality closely interwoven: first, formulating cognitive frameworks concerning the goals, means and environment of action; second, activating relationships among the actors, who communicate, negotiate, and make decisions; and third, making emotional investments, which enable individuals to recognize themselves in each other" (p. 35). All this is interesting, but it seems to inflate the idea of collective identity beyond useful proportion.

6. Elsewhere, the first author has used the term "collective identity work" to mean what is meant here by subcultural identity work. In that other text (Schwalbe 1996), written for a wider audience, the risk of confusion with the social movements notion of collective identity seemed slight.

7. Collas (1994) discusses this phenomenon with regard to racial and ethic identities, Walker (1993) with regard to sexual identities. Barrie Thorne (1993) has also shown how adults encourage children to create the boundaries that reproduce and maintain gender identities.

8. A moral identity is one that implies special virtue or wickedness and can be claimed or imposed only if strict criteria are met. For example, in our culture, *mother* is a moral identity available only to women who bear children. *Breadwinner* is a moral identity that was once the exclusive claim of men who were family-supporting workers. *Convict* is also a moral identity, though not usually a desirable one, that requires strict tests prior to imposition. As we say in the text, any identity can be made into a moral identity. What matters is how it is defined within a particular community.

9. See Gergen (1991) for an exception and Dowd (1991) for a different take on the matter.

REFERENCES

Alexander, C. N., and Mary G. Wiley. 1981. "Situated Activity and Identity Formation." Pp. 269-289 in *Social Psychology: Sociological Perspectives*, edited by M. Rosenberg and R. Turner. New York: Basic.
Allen, T. 1994. *The Invention of the White Race*. London: Verso.
Anderson, E. 1992. *StreetWise*. Chicago: University of Chicago Press.
Baumeister, R. 1982. "A Self-Presentational View of Social Phenomena." *Psychological Bulletin* 91: 3-26.
Bem, S. 1981. "Gender Schema Theory: A Cognitive Account of Sex Typing." *Psychological Review* 88: 369-371.
Blumer, H. 1969. *Symbolic Interactionism*. Englewood Cliffs, NJ: Prentice-Hall.
Brisset, D., and C. Edgley. Eds. 1990. *Life as Theater*, 2nd ed. New York: Aldine.
Burke, K. 1957. *The Philosophy of Literary Form*. New York: Vintage.
Burke, P., and D. C. Reitzes. 1981. "The Link Between Identity and Role Performance." *Social Psychology Quarterly* 44: 83-92.
Calhoun, C. Ed. 1994. *Social Theory and the Politics of Identity*. New York: Blackwell.
Charmaz, K. 1987. "Struggling for a Self: Identity Levels of the Chronically Ill." Pp. 283-321 in *Research in the Sociology of Health Care: The Experience and Management of Chronic Illness*, Vol. 6, edited by J.A. Roth and P. Conrad. Greenwich, CT: JAI Press.
Collas, S. 1994. *Uncle Tom and Oreo: The Debates of "Authenticity" in Racial and Ethnic Identities*. Paper presented at annual meetings of the American Sociological Association, August, Los Angeles, CA.
Davis, F. 1992. *Fashion, Culture, and Identity*. Chicago: University of Chicago Press.

Denzin, N. 1988. "Act, Language and Self in Symbolic Interactionist Thought." *Studies in Symbolic Interaction* 9: 51-80.

Deschamps, J.C. 1982. "Social Identity and Relations of Power Between Groups." Pp. 85-98 in *Social Identity and Intergroup Relations*, edited by H. Tajfel. Cambridge, UK: Cambridge University Press.

Dowd, J. 1991. "Social Psychology in a Postmodern Age: A Discipline Without a Subject." *The American Sociologist* (Fall/Winter): 188-209.

Farr, R. M., and S. Moscovici. Eds. 1984. *Social Representations*. New York: Cambridge Unitersity Press.

Fields, B. 1990. "Slavery, Race, and Ideology in the United States of America." *New Left Review* 179: 95-117.

Fine, G. A., and S. Kleinman. 1979. "Rethinking Subculture: An Interactionist Analysis." *American Journal of Sociology* 85: 1-20.

Fine, M. 1994. "Working the Hyphens: Reinventing Self and Other in Qualitative Research." Pp. 70-82 in *Handbook of Qualitative Research*, edited by N. Denzin and Y. Lincoln. Thousand Oaks, CA: Sage.

Finkelstein, J. 1991. *The Fashioned Self*. Philadelphia: Temple University Press.

Fordham, S., and J. Ogbu. 1986. "Black Students' School Success: Coping With the 'Burden of Acting White.'" *Urban Review* 18: 176-206.

Frankenberg, R. 1993. *White Women, Race Matters: The Social Construction of Whiteness*. Minneapolis: University of Minnesota Press.

Friedman, D., and D. McAdam. 1992. "Identity Incentives and Activism: Networks, Choices, and the Life of a Social Movement." Pp. 156-173 in *Frontiers in Social Movement Theory*, edited by A. Morris and C. Mueller. New Haven, CT: Yale University Press.

Gecas, V., and P. Burke. 1995. "Self and Identity." Pp. 41-67 in *Sociological Perspectives on Social Psychology*, edited by K. Cook, G.A. Fine, and J. House. Boston: Allyn and Bacon.

Gergen, K. 1991. *The Saturated Self*. New York: Basic.

Gergen, K., and M. Gergen. 1983. "Narratives of the Self." Pp. 254-273 in *Studies in Social Identity*, edited by T.R. Sarbin and K.E. Scheibe. New York: Praeger.

Goffman, E. 1959. *The Presentation of Self in Everyday Life*. New York: Doubleday.

_____. 1961. *Encounters: Two Studies in the Sociology of Interaction*. Indianapolis: Bobbs-Merrill.

_____. 1963. *Stigma: Notes on the Management of Spoiled Identity*. Englewood Cliffs, NJ: Prentice-Hall.

_____. 1967. *Interaction Ritual*. New York: Doubleday.

_____. 1983. "The Interaction Order." *American Sociological Review* 48: 1-17.

Gordon, C. 1968. "Self-Conceptions: Configurations of Content." Pp. 115-136 in *The Self in Social Interaction*, Vol. 1, edited by C. Gordon and K. Gergen. New York: John Wiley.

Hadden, S., and M. Lester. 1978. "Talking Identity: The Production of 'Self' in Interaction." *Human Studies* 1: 331-356.

Hewitt, J. 1989. *Dilemmas of the American Self*. Philadelphia: Temple University Press.

Hewitt, J., and R. Stokes. 1975. "Disclaimers." *American Sociological Review* 40: 1-11.

Hunt, S., and R. Benford. 1994. "Identity Talk in the Peace and Justice Movement." *Journal of Contemporary Ethnography* 22: 488-517.

Hunt, S., R. Benford, and D. Snow. 1994. "Identity Fields: Framing Processes and the Social Construction of Movement Identities." Pp. 185-208 in *New Social Movements: From Ideology to Identity*, edited by E. Larana, H. Johnston, and J. Gusfield. Philadephia: Temple University Press.

Jones, E. E., and T. S. Pittman. 1982. "Toward a General Theory of Strategic Self-Presentation." Pp. 231-262 in *Psychological Perspectives on the Self*, Vol. 1, edited by J. Suls. Hillsdale, NJ: Lawrence Erlbaum.

Kessler, S., and W. McKenna. 1978. *Gender: An Ethnomethodological Approach*. Chicago: University of Chicago Press.

Kitzinger, C. 1989. "Liberal Humanism as an Ideology of Social Control: The Regulation of Lesbian Identities." Pp. 82-98 in *Texts of Identity*, edited by J. Shotter and K. Gergen. Newbury Park, CA: Sage.

Klapp, O. 1969. *Collective Search for Identity.* New York: Holt, Rinehart and Winston.

Lash, S., and J. Friedman. Eds. 1992. *Modernity and Identity.* Cambridge, MA: Basil Blackwell.

Lofland, J. 1969. *Deviance and Identity.* Englewood Cliffs, NJ: Prentice-Hall.

Lyman, S., and M. Scott. 1968. "Accounts." *American Sociological Review* 33: 46-62.

MacCannell, D. 1976. "The Past and Future of Symbolic Interactionism." *Semiotica* 16: 99-114.

MacLeod, J. 1987. *Ain't No Making It.* Boulder, CO: Westview.

Majors, R., and J. M. Billson. 1992. *Cool Pose.* New York: Lexington.

Manning, P. K. 1987. *Semiotics and Fieldwork* (Qualitative Research Methods, Vol. 7). Beverly Hills, CA: Sage.

Manning, P. K., and B. Cullum-Swan. 1994. "Narrative, Content, and Semiotic Analysis." Pp. 463-477 in *Handbook of Qualitative Research*, edited by N. Denzin and Y. Lincoln. Thousand Oaks, CA: Sage.

Mason-Schrock, D. P. 1995. "The Identity Work of Transsexuals." Master's thesis. Department of Sociology, North Carolina State University.

Melucci, A. 1989. *Nomads of the Present: Social Movements and Individual Needs in Contemporary Society.* Philadelphia: Temple University Press.

McCall, G., and J. L. Simmons. 1978. *Identities and Interactions,* rev. ed. New York: The Free Press.

Mol, H. 1976. *Identity and the Sacred.* New York: The Free Press.

Nagel, J. 1994. "Constructing Ethnicity: Creating and Recreating Ethnic Identity and Culture." *Social Problems* 41: 152-176.

Ogbu, J. 1978. *Minority Education and Caste: The American System in Cross-Cultural Perspective.* New York: Academic Press.

Omi, M., and H. Winant. 1994. *Racial Formation in the United States: From the 1960s to the 1980s,* 2nd ed. New York: Routledge.

Perinbanayagam, R. S. 1985. *Signifying Acts.* Carbondale, IL: Southern Illinois University Press.

———. 1991. *Discursive Acts.* New York: Aldine.

Rochberg-Halton, E. 1982. "Situation, Structure, and the Context of Meaning." *Sociological Quarterly* 23: 455-476.

Roediger, D. R. 1991. *The Wages of Whiteness: Race and the Making of the American Working Class.* London: Verso.

Sandstrom, K. 1990. "Confronting Deadly Disease: The Drama of Identity Construction Among Gay Men With AIDS." *Journal of Contemporary Ethnography* 19: 271-294.

Saxton, A. 1990. *The Rise and Fall of the White Republic: Class Politics and Mass Culture in Nineteenth-Century America.* London: Verso.

Schlenker, B. 1980. *Impression Management.* Monterey, CA: Brooks/Cole.

Schwalbe, M. L. 1983. "Language and the Self." *Symbolic Interaction* 6: 291-306.

———. 1993. "Goffman Against Post-modernism: Emotion and the Reality of the Self." *Symbolic Interaction* 16: 333-350.

———. 1995. "Mythopoetic Men's Work as a Search for Communitas." Pp. 507-519 in *Men's Lives*, 3rd ed., edited by M. Kimmel and M. Messner. Boston: Allyn and Bacon.

———. 1996. *Unlocking the Iron Cage: The Men's Movement, Gender Politics, and American Culture.* New York: Oxford.

Sebeok, T. 1979. "The Semiotic Self." Pp. 263-267 in *The Sign and Its Masters*, edited by T. Sebeok. Austin: University of Texas Press.

Sennett, R., and J. Cobb. 1972. *The Hidden Injuries of Class.* New York: Vintage.

Singer, M. 1980. "Signs of the Self: An Exploration in Semiotic Anthropology." *American Anthropologist* 82: 485-507.

Snow, D., and L. Anderson. 1987. "Identity Work Among the Homeless: The Verbal Construction and Avowal of Personal Identities." *American Journal of Sociology* 92: 1336-1371.

Staples, R. 1978. "Masculinity and Race: The Dual Dilemma of Black Men." *Journal of Social Issues* 34: 169-183.

Stone, G. P. 1981. "Appearance and the Self: A Slightly Revised Wersion." Pp. 187-202 in *Social Psychology Through Symbolic Interactionism*, edited by G. Stone and H. Farberman. New York: John Wiley.

Strauss, A. 1959. *Mirrors and Masks*. Glencoe, IL: The Free Press.

Stryker, S. 1968. "Identity Salience and Role Performance." *Journal of Marriage and the Family* 30: 558-564.

_____. 1980. *Symbolic Interactionism: A Social Structural Version*. Menlo Park, CA: Benjamin/ Cummings.

Tajfel, H. 1981. *Human Groups and Social Categories: Studies in Social Psychology*. Cambridge, UK: Cambridge University Press.

Taylor, C. 1989. *Sources of the Self*. Cambridge, MA: Harvard University Press.

Taylor, V., and N. E. Whittier. 1992. "Collective Identity in Social Movement Communities." Pp. 104-129 in *Frontiers in Social Movement Theory*, edited by A. Morris and C. Mueller. New Haven, CT: Yale University Press.

Tedeschi, J. Ed. 1981. *Impression Management Theory and Social Psychological Research*. New York: Academic Press.

Thorne, B. 1993. *Gender Play: Boys and Girls in School*. New Brunswick, NJ: Rutgers University Press.

Thumma, S. 1991. "Negotiating a Religious Identity: The Case of the Gay Evangelical." *Sociological Analysis* 52: 333-347.

Tolson, A. 1977. *The Limits of Masculinity: Male Identity and Women's Liberation*. New York: Harper and Row.

Walker, L. M. 1993. "How to Recognize a Lesbian: The Cultural Politics of Looking Like What You Are." *Signs* 18: 866-890.

Weigert, A., J. S. Teitge, and D. Teitge. 1986. *Society and Identity*. New York: Cambridge University Press.

Weinstein, E., and P. Deutschberger. 1963. "Some Dimensions of Altercasting." *Sociometry* 26: 454-456.

West, C., and D. H. Zimmerman. 1987. "Doing Gender." *Gender & Society* 1: 125-151.

Wicklund, R. A., and P. M. Gollwitzer. 1982. *Symbolic Self-Completion*. Hillsdale, NJ: Lawrence Erlbaum.

Wieder, L. D. 1974. *Language and Social Reality*. The Hague: Mouton.

Willis, P. 1977. *Learning to Labor*. Aldershot, UK: Gower.

Young, K. 1989. "Narrative Embodiments: Enclaves of the Self in the Realm of Medicine." Pp. 152-165 in *Texts of Identity*, edited by J. Shotter and K. Gergen. Newbury Park, CA: Sage.

Young, T. R. 1990. *The Drama of Social Life*. New Brunswick, NJ: Transaction.

PUNISHMENT AND COERCION IN SOCIAL EXCHANGE

Linda D. Molm

ABSTRACT

Social exchange theorists have traditionally excluded punishment and coercion from the scope of their analyses. This chapter summarizes a decade-long program of theory and research that sought to incorporate punishment and coercion in exchange theory, particularly power-dependence theory, and to explain the use and effects of coercive power in social exchange relations. Three main arguments are supported: First, unlike reward power, the use of coercive power is not structurally induced by power advantage; that is, the power to punish does not lead automatically to its use. Coercion is a purposive power strategy in which actors knowingly and contingently impose cost on another to compel them to provide rewards. Second, in relations in which actors have the power to both reward and punish each other, the use of coercion is risky. Reward dependence provides the incentive to coerce, but it also increases the costs that a partner can impose in retaliation. Justice norms reinforce these effects. Consequently, loss averse actors typically forego coercion and its potential gains to avoid loss; without use, coercive power becomes ineffective. Third, when used consistently and contingently, coercion is a powerful means of obtaining rewards. In their effects on exchange, coercive power and reward power follow the same principles.

Advances in Group Processes, Volume 13, pages 149-188.
ISBN: 0-7623-0005-1

This chapter reviews the development of a theoretical research program on coercive power in social exchange. Conducted within the framework of Emerson's theory of power-dependence relations, the decade-long program sought to extend the theory by incorporating punishment and coercion and to explain the use and effects of coercive power in social exchange relations. Its focus was exchange relations of some endurance that are imbedded in larger networks, formed and maintained without explicit bargaining or negotiation, and performed by actors who control both rewarding and punishing outcomes for one another.

The original question that this program addressed was whether reward power and coercive power, when equivalent on all dimensions but sign of outcome (positive or negative), would have comparable effects on social exchange. Blau (1964) and Homans (1974), both of whom excluded punishment and coercion from the scope of their exchange theories, argued that they would not have the same effects: that punishment would lead to hostility, retaliation, and the destruction of mutually beneficial exchange. They based this argument primarily on the erroneous conclusions of early behavioral research on punishment (e.g., Estes 1944; Skinner 1938; Thorndike 1932).[1] Many social scientists share the view that coercion is ineffective and destructive, however, including political theorists (e.g., Boulding 1969; Etzioni 1968), bargaining researchers (e.g., Lawler 1992; Pruitt 1981; Rubin and Brown 1975), and gaming theorists (e.g., Deutsch 1973; Tedeschi, Schlenker, and Bonoma 1973). Most of these theories assume that punishment provokes retaliation, which in return leads to counter-retaliation and the escalation of mutual conflict.

Heath (1976), in contrast, argued that mutual reward exchanges and coercive exchanges are fundamentally the same and that both could be explained by exchange principles. Actors give rewards to another in exchange for something they value, either reciprocal rewards or the removal of punishment or threat of harm. In mutual reward exchanges, actors give rewards in exchange for reciprocal rewards; in coercive exchanges, one actor gives rewards in exchange for the other's cessation or withholding of punishment. Consequently, actors who control either rewarding or punishing outcomes for others have power over them. Far from being ineffective, some theorists (e.g., Anderson and Willer 1981; Wrong 1979) argue that coercive power is highly effective.

As my research shows, neither position fully describes the complexity of the differences between reward power and coercive power. Both are partially correct. Early experiments in the program left little doubt that the two bases of power do *not* have equivalent effects on the frequency and distribution of exchange benefits in relations in which actors control both rewards and punishments for one another. Instead, the patterns of exchange that develop are almost entirely a function of the structure of reward power. Coercive power has little effect of any kind, positive or negative.

Rather than refuting Heath's position, however, the later work in the program developed and tested a theory that, in some respects, supports it. This theory offers three main conclusions about the similarities and differences between reward power and coercive power. First, when coercive power is consistently used, no new principles are required to explain its effects on exchange. As Heath proposed, actors will increase the frequency of their reward exchange either to obtain rewards or to avoid punishments. If anything, actors' aversion to loss enhances the effects of coercive power so that it is more potent than reward power. Second, one of the central tenets of power-dependence theory holds only for reward power. The theory predicts that power advantage leads to power use, regardless of intent to use power or awareness of power. But a structural advantage on coercive power does not induce the use of that power. The primary distinction between reward power and coercive power lies not in their *effects* on social exchange, as Blau and Homans proposed, but in the relation between the *structure* and *use* of power. Third, extending exchange theory's analysis of power-dependence relations to coercive power requires developing predictions about strategic power use. Coercive power use is not an unintended byproduct of actors seeking their own self-interest; rather, it is a purposive strategy for increasing an exchange partner's rewards through the contingent application of punishment. Because the structural condition that provides the incentive to use coercion—high reward dependence—also increases the risk of loss, coercion is constrained. Rather than opposing these structural effects, justice norms support and maintain them.

In this chapter, I describe the central concepts and assumptions of the theory, the standard experimental setting in which the theory was developed and tested, and the theoretical derivations and research findings that underly the three main conclusions.

BASIC CONCEPTS AND
ASSUMPTIONS OF SOCIAL EXCHANGE

The research program adopts and builds upon the social exchange theory formulated by Richard Emerson (1972a, 1972b, 1981), particularly his analysis of power-dependence relations. It also reflects the emphasis of Blau and Homans on nonnegotiated social exchanges, draws on Thibaut and Kelley's (1959) analysis of power strategies, and incorporates Kahneman and Tversky's (1979) assumptions about the value function of outcomes.

Emerson's theory applies to forms of social interaction that meet five general scope conditions: (1) actors behave in ways that increase outcomes they positively value and decrease outcomes they negatively value, (2) classes of valued outcomes obey a principle of satiation (in psychological terms) or declining marginal utility (in economic terms), (3) the structure of the relation between actors is one of mutual dependence, (4) benefits obtained from others are contingent on benefits given "in exchange," and (5) exchanges of benefits between the same actors or sets of actors are recurring over time (Emerson 1981; Molm and Cook 1995).[2]

The first condition states the behavioral assumption of self-interest that underlies all theories of social exchange, whether based on psychological principles of learning or economic principles of rational choice. The second restricts the theory to outcomes whose value diminishes as more are received (an assumption shared, again, by most behavioral and economic theories). The third describes the basic structure of the social relations to which the theory applies; that is, actors exchange benefits only if each is dependent, to some degree, on the other for obtaining those benefits. The fourth describes the process that gives the theory its name and limits its scope to social behaviors that are contingent on their returns. And the fifth restricts the theory to continuing exchange relations with a history and a future; that is, "one-shot" transactions are excluded from its scope. I adopt these scope conditions and, in addition, restrict my analysis to several other conditions discussed in the following sections.

The basic concepts and assumptions of the theory are organized around four main elements: (1) the actors who exchange, (2) the exchange resources, (3) the structure of exchange, and (4) the process of exchange.

Actors

Social exchange minimally involves two or more *actors*, the entities who engage in social exchange. Actors can be either individuals or corporate groups acting as a single unit and either specific entities or interchangeable occupants of structural positions. Within the general scope condition of self-interested actors, the following assumptions state the specific learning principles on which Emerson's (1972a) original formulation was based and that underly the theory's structural predictions:

Assumption 1.1.   Actors initiate exchanges by giving rewards to other actors who control resources they value.

Assumption 1.2.   Actors increase patterns of exchange that are relatively more rewarding (or less punishing), decrease those that are relatively less rewarding (or more punishing), and change behaviors (i.e., engage in exploratory behavior) when rewards decline or punishments increase.

Assumption 1.1 provides the necessary bridge from theories of individual behavior to social exchange; that is, in relations governed by the fourth scope condition, actors learn to initiate exchange with those who control rewards they value. These initiations then increase or decrease according to their consequences (Assumption 1.2).

Resources, Outcomes, and Value

Several concepts refer to aspects of what actors exchange: resources, outcomes, rewards, punishments, costs, value, and exchange domains. An actor's capacity to provide some benefit for another is a *resource* in that relation. Resources include both tangible possessions that can be transferred to another and behavioral capabilities to produce value for another (Emerson 1972b, 1981). When an actor provides benefit for another, she incurs *cost*. These costs can include one or more of the following: opportunity costs, investment costs, the actual loss of a material resource, or costs intrinsic to the behavior itself (e.g., fatigue or pain). My analysis is restricted to exchanges in which (1) resources are capacities to perform behaviors that produce valued outcomes for others, rather than tangible goods, and (2) the only costs to the actor are opportunity costs.[3]

The *outcomes* of exchange refer to the rewards or punishments that actors receive from each other in exchange. Outcomes can have positive value (*rewards*) or negative value (*punishments*). Assumption 2 specifies three properties of the value function for outcomes:

Assumption 2. The value function for outcomes from reciprocal exchange is described by the properties of referent dependence, diminishing sensitivity, and loss aversion.

 a. *Referent dependence.* Outcomes are evaluated in relation to a reference point, the status quo. Outcomes count as *gains* or *rewards* if they improve an actor's current outcome level in an exchange relation, and as *losses* or *punishments* if they worsen it.
 b. *Diminishing sensitivity.* The marginal value of gains and losses decreases with their distance from the reference point.
 c. *Loss aversion.* The negative subjective value of a loss is greater than the positive subjective value of an equivalent gain.

These properties of the value function are most commonly associated with Kahneman and Tversky's (1979) prospect theory, which is based on empirical generalizations derived from studies of decision making under risk and uncertainty—conditions which also characterize the exchange relations I study. As Tversky and Kahneman note, however, they are equally valid for describing conditioned reactions to consequences of past behavior: "organisms habituate to steady states, the marginal response to changes is diminishing, and pain is more urgent than pleasure" (1991, p. 1057).

The first property, referent dependence, states that outcomes have no a priori status as rewards or punishers; whether a particular outcome is rewarding or punishing depends on its relation to an actor's current situation. This conception of rewards and punishments as referent dependent, with the status quo as the typical reference point, is common to numerous theories that posit some process of value adaptation, including contemporary behavioral theory (Van Houten 1983) and Thibaut and Kelley's (1959) analysis of the comparison level (CL) of exchange relations. An important implication of this property is that rewards and punishments are potentially interchangeable: if rewards are regularly given, they can be withheld as punishment, and if punishments are regularly given, they can be withheld as rewards (Bacharach and Lawler 1981; Blau 1964). This assumption is central to the concept of coerced exchange.

The second property, diminishing sensitivity, corresponds to the scope condition that all valued outcomes obey a principle of satiation or diminishing marginal utility. The only difference lies in scaling outcomes from a reference point (Kahneman and Tversky 1979). Emerson used this principle to define an *exchange domain* (Emerson 1972a). Outcomes belong to a single class or domain if the receipt of one outcome reduces the value of all outcomes within that domain. While the value of benefits can vary both across domains and within domains, my analysis, like most exchange analyses, is restricted to a single exchange domain. It considers variations in value within that domain and defines rewards and punishments as gains or losses in value.

The third property, loss aversion, describes the finding of numerous studies by economists, psychologists, and sociologists (e.g., Fishburn and Kochenberger 1979; Gray and Tallman 1987; Hershey and Schoemaker 1980; Kahneman and Tversky 1979, 1982, 1984; Tversky and Kahneman 1991). When faced with choices between behaviors with uncertain outcomes, actors weigh potential losses more heavily than potential gains. Similarly, behavioral studies show that punishment suppresses behavior faster and to a greater extent than reinforcement strengthens it (see Patterson 1982; Van Houten 1983 for reviews). Studies typically report coefficients of loss aversion in the range of 2 to 2.5, indicating that the slope of the value function for losses is roughly twice the slope for gains (Kahneman, Knetsch, and Thaler 1991; Molm 1991; Tversky and Kahneman 1991).[4]

Exchange Structure

As the third scope condition states, social exchange relations develop within structures of mutual dependence. An actor is *dependent* on another for outcomes within an exchange domain to the extent that those outcomes are contingent on exchange with the other. Dependence varies with value and alternatives (Emerson 1972a):

Assumption 3.   *A*'s dependence on *B* varies directly with the *value* of the outcomes that *B* can produce for *A* and inversely with *A*'s *alternative* sources of outcomes within the same domain.

I extend the concept of dependence to negative outcomes; that is, actors can be dependent on one another for obtaining positively valued outcomes or for avoiding negatively valued outcomes. If *B*'s behavior can produce outcomes for *A* that are either positive or negative in value, then *A*'s total dependence on *B* equals the *range* of outcomes through which *B*'s behavior can move *A*. For the purpose of comparing these two forms of dependence, I define and operationalize each separately.

A's alternatives to exchange with *B* are determined by the structure of the *exchange network* within which the *A—B* relation is embedded. An exchange network is a set of two or more *connected* dyadic exchange relations (Emerson 1972b). Relations are connected if the frequency of exchange in one relation affects the frequency of exchange in another. For example, in the simple network *B—A—C*, the *B—A* and *A—C* relations are connected if the frequency of exchange between *A* and *B* affects the frequency of exchange between *A* and *C*. Connections are *positive* to the extent that exchange in one relation increases exchange in the other, *negative* to the extent that exchange in one relation decreases exchange in the other, and *null* if exchange in one relation has no effect on exchange in another (i.e., the two relations are independent).

Negative connections between two (or more) relations (e.g., *B—A* and *A—C*) typically occur when *B* and *C* both control resources that offer value for *A* in the same exchange domain (Emerson 1972b; Yamaguchi 1996). For example, *B* and *C* might offer the same kind of advice, support, or enjoyment. Then, *B* and *C* provide *A* with *alternative* opportunities for obtaining value in that domain, and they are competitors for the resources that *A* controls. Positive connections are created when exchange domains are not only different, but the outcome of *A*'s exchange with *B* is a valuable resource for *A* in *A*'s exchange with another actor, *C*. Then, the one exchange facilitates, rather than hinders, the other.

Although I study power in exchange networks, my unit of analysis is the dyadic exchange relation. My interest is in the power relations and exchange processes within the dyad, not the distribution of power in the network as a whole. I limit the scope of my analysis to the following structural conditions: (1) exchange relations are embedded in larger, negatively-connected networks (i.e., all actors have alternative partners who can provide resources in the same domain); (2) all actors are dependent on one another for obtaining rewards *and* avoiding punishments (i.e., all actors have the capacity to *both* reward and punish each other); (3) actors cannot change the structure of their relations, nor can they leave the field of influence (i.e., they cannot avoid the rewarding or punishing actions of others to whom they are connected in the network).

Exchange Process

An *exchange opportunity* is a situation or event that provides actors with the occasion to *initiate* an exchange by providing a benefit for another actor. When an initiation is reciprocated, the mutual exchange of benefits that results is called a *transaction*. An ongoing series of transactions between the same two actors constitutes an *exchange relation*.

The fourth and fifth scope conditions of the theory refer to the process of exchange. The fourth condition, that "benefits obtained ... are contingent upon benefits provided" (Emerson 1981, p. 31), makes some degree of reciprocity a defining condition of social exchange. Reciprocity need not be equal or immediate, however. The fifth condition, that actors engage in recurring transactions with specific partners over time, extends the condition of contingency *within* transactions to a requirement of contingency *between* transactions. The interdependence of sequential transactions provides the opportunity for actors to influence their partners' behaviors in ways that are impossible when transactions are independent (Molm 1994b).

In relations of direct exchange, exchange transactions can take one of two forms: reciprocal and negotiated (Emerson 1981). In *negotiated exchange*, actors engage in a joint decision process, such as explicit bargaining, in which they agree on the terms of an exchange that provides known outcomes for both. Exchange relations take the form of a series of discrete, two-sided transactions. In *reciprocal exchange*, actors individually provide benefit for each other without negotiation and without knowing whether, when, or to what degree the other will reciprocate. The relation takes the form of a series of individually performed, sequentially contingent acts, variable in reciprocity and timing. In contrast to many contemporary theories of network exchange (e.g., Cook and Yamagishi 1992; Friedkin 1993; Skvoretz and Willer 1993), which are restricted to negotiated transactions, my analysis is restricted to reciprocal transactions.

In relations in which actors can both reward and punish each other, a reciprocal exchange relation is likely, over time, to include components of three "ideal types" of transaction: mutual reward, mutual punishment, and coercion. In *mutual reward* exchanges, actors engage in a mutually contingent (but not necessarily equal) exchange of rewards. In *mutual punishment*, actors engage in a mutually contingent exchange of punishments, often described as *conflict* (e.g., Blalock 1989; Tedeschi, Schlenker, and Bonoma 1973; Willer 1981). I will refer to the contingency of one actor's punishment on another's prior punishment as *retaliatory punishment*. *Coercion*, in contrast to either mutual reward or mutual punishment, involves the potential flow of rewards from one actor and punishment from another (Willer 1981). A coerces B by punishing B when B fails to reward A and withholding punishment when B's rewards to A are forthcoming. A variation on this pattern of "pure" coercion, and one which I study in this program, is when A coerces B by punishing B's failure to reward but reciprocates B's rewards (and withholds punishment) when they are forthcoming. Such a pattern explicitly requires control over *both* rewards and punishments for another actor.

In both mutual reward exchanges and coercive exchanges, actors exchange benefits. In coercive exchanges, however, the "benefit" for the target of coercion is a reduction or cessation of punishment. Such action is rewarding only if the coercer first establishes the expectation that punishment will be imposed, through either threats or a history of contingent punishment, if the other fails to provide rewards. Thus, a coercive exchange consists of an exchange of a reward for the withholding of expected punishment.[5]

POWER AND POWER USE

Structural Power

The mutual dependence of actors on one another for valued outcomes provides the structural basis for their power over each other (Emerson 1962, 1972b). A's *power* over B (P_{ab}) is defined as the level of potential cost that A can impose on B. It derives from, and is equal to, B's dependence on A (D_{ba}) (Emerson 1972b):

Assumption 4. A's power over B equals B's dependence on A; that is, $P_{ab} = D_{ba}$.

If A and B are equally dependent on each other, power in the relation is *balanced*. If their dependencies are unequal, power is *imbalanced*. The less dependent and more powerful actor has a structural *power advantage* in the relation, and the more dependent actor is *power disadvantaged*.

Because each actor's power is a function of the other's dependence, the power relation between two actors in a relation is not zero-sum (Lawler 1992). Two dimensions of relational power—the absolute power in a relation, and the relative power in a relation—can be independently defined. The *average power* in a relation, defined as the average of two actors' dependencies on each other, is a measure of the absolute strength of the actors' power over each other. *Power imbalance*, defined as the difference between two actors' dependencies on each other, is a measure of their relative power over each other.

If the concept of dependence is extended to include dependence on another who controls negative as well as positive outcomes, then—by Assumption 4—power is also extended to include both reward-based and punishment-based power. A's *reward power* over B is equal to B's dependence on A for obtaining rewards, and A's *punishment power* over B is equal to B's dependence on A for avoiding punishment. Defining the average power and power imbalance for each of these bases of power produces four central variables that describe the structural power in a dyadic relation: average reward power, average punishment power, reward power imbalance, and punishment power imbalance.

Power Use

Whereas power imbalance and average power are structural attributes of exchange relations that determine actors' potential power over each other, *power use* is the behavioral exercise of that potential in interaction. Actors use power, of either base, by imposing cost on the partner. The amount of cost they can impose is equal to the other's dependence (Emerson 1972b). In balanced relations, both actors can impose equal cost on each other. But in imbalanced relations, actors who are power-advantaged can impose greater costs than their more dependent partners. Over time, the structure of power and dependence produces predictable effects on the frequency and distribution of exchange as actors use power to maintain exchange or gain advantage.

Three theorems about the relation between structural power and behavioral exchange can be derived from the previous assumptions:

Theorem 1. A's initiations of exchange with B increase with A's dependence on B.

Theorem 2. The average frequency of exchange in a relation increases with average power.

Theorem 3. The asymmetry of exchange in a relation increases with power imbalance, in favor of the more powerful actor. If A has a power advantage, then A's use of power will increase over time, as evidenced by either (a) increased rewards to A from B (i.e., increased costs for B), or (b) decreased rewards to B from A (i.e., decreased costs for A).[6]

Numerous studies support these predictions. Across a variety of exchange settings, research shows that the average frequency of exchange in a relation increases with the average power of the relation (e.g., Bacharach and Lawler 1981; Lawler and Bacharach 1987; Michaels and Wiggins 1976), and the asymmetry of reward exchange increases as power imbalance increases, in favor of the power-advantaged actor (e.g., Burgess and Nielsen 1974; Cook and Emerson 1978; Cook, Emerson, Gillmore, and Yamagishi 1983; Markovsky, Willer, and Patton 1988; Molm 1981, 1985).[7] Power imbalance also affects the average frequency of exchange; holding average power constant, the frequency with which actors exchange tends to decrease as power imbalance increases (e.g., Molm 1990; Lawler and Yoon 1996).

In this research program, I examine whether these predictions hold for punishment power. If they do, then A should be more likely to offer rewards to actors on whom A is more dependent for either obtaining rewards or avoiding punishments (Theorem 1), the frequency of reward exchange in a relation should increase with either average reward power or average punishment power (Theorem 2), and the asymmetry of reward exchange (i.e., power use) should increase with either reward power imbalance or punishment power imbalance, in favor of the actor who is less dependent and power-advantaged (Theorem 3).

Structurally, the two bases of power offer the same potential for power use. Control over either rewards or punishments for others gives actors the means to impose cost on another. When actors use reward power, they do so by withholding benefits from the partner; when they use punishment power, they inflict actual losses on the partner (Bacharach and Lawler 1981). The costs they impose in the long term, however, are the same: if effective, both forms of power ultimately induce the disadvantaged partner in an imbalanced relation to accept greater opportunity costs in exchange for either rewards or the cessation of punishment. The disadvantaged actor forgoes potential benefits from alternative relations to provide rewards for the advantaged partner, while the advantaged partner either reciprocates the rewards—but less frequently (reward power), or withholds punishment (punishment power).

Structurally-Induced or Strategic Power Use

The use of power can be *structurally induced* by power advantage, regardless of actors' intent to use power or to influence another's behavior, or it can be *strategic*. Emerson (1972b) allowed for both forms of power use, but argued that a conception of "power use" as a distinct mode of voluntary action was unnecessary. Regardless of actors' intentions or their use of behavioral influence strategies, a structural power advantage would lead to power use; that is, "to have a power advantage is to *use* it" (Emerson 1972b, p. 67, his emphasis).

The mechanism that drives this process is the structure of the negatively-connected network, which provides power-advantaged actors with access to more or better alternatives. When these actors pursue exchange with their alternatives (Assumption 1.1), they inadvertently withhold rewards from their more dependent partners; that is, they "use power" over them. In the process, they drive up the cost of obtaining the rewards they control while lowering their own cost of obtaining their partners' rewards. As long as actors follow the behavioral principles of Assumption 1.2, a power imbalanced structure will produce the distribution of exchange predicted in Theorem 3. Whether punishment power affects behavior through a comparable process was the first issue addressed in this program.

Actors can also use power strategically, by selectively giving and withholding rewards or punishments, contingent on the partner's prior behavior. This is the form of power use that Thibaut and Kelley (1959) assumed when they discussed the conversion of fate control (structural dependence) to behavior control. When actors use power strategically, they *create* contingencies that produce predictable consequences for their partners' behaviors, rather than simply responding to the consequences of their own behaviors. They make their rewarding actions contingent on some level of prior rewarding by the exchange partner, or they contingently punish the other's failure to provide sufficient rewards.

Strategic power use is not structurally determined, although the structure of power can produce exchange outcomes that motivate the use of contingent influence strategies.[9] Structure also sets the boundaries within which actors can influence exchange ratios through strategic power use by determining the best outcomes that actors in different power positions can hope to obtain (Michaels and Wiggins 1976). And, as I discuss, structure affects the consequences of strategic power use, which in turn influence whether it continues or declines.

Contingent influence strategies, which I call *power strategies*, are only possible in relations in which the same actors repeatedly exchange with one another (the fifth scope condition). To facilitate the analysis of purposive power strategies, I also assume that actors have full information about the values governing relative dependencies in their immediate exchange relations. That condition makes it possible for actors to use purposive power strategies that take account of their relative power in the relation.

THE STANDARDIZED LABORATORY SETTING

The research was conducted in a standardized laboratory setting designed to meet the general scope conditions of power-dependence theory and the additional conditions assumed for my analysis. In all experiments, subjects earned money through reciprocal exchange with partners to whom they were connected in a four-actor exchange network. Subjects were mutually dependent on one another for their monetary earnings in the experiment, and they exchanged points—worth money—via computer while seated in isolated rooms. I manipulated dimensions of power by varying the amounts of money that actors in the network could add to (reward power) or subtract from (punishment power) their partners' earnings on each of a series of exchange opportunities. Subjects engaged in reciprocal, nonnegotiated exchanges (i.e., they individually made choices on their computers that affected other actors' earnings) with the same partners over repeated opportunities. The frequencies and contingencies of their rewarding and punishing actions toward each other across these opportunities provided the data for computing measures of power use and coercion.

Operationalization of Exchange Concepts

Actors

Actors were undergraduate student subjects recruited on the basis of their desire to earn money. Subjects were randomly assigned to positions in networks, and networks to experimental conditions. All experiments were conducted with 10 networks per condition (and with equal numbers of all-male and all-female networks); thus, all tests of significance were based on the same number of cases within conditions.

Resources

Money is the valued outcome in the experiments, and actors' resources are their capacities to perform behaviors that produce monetary gains or losses for another. In contrast to economic exchanges, money was not transferred from one actor to another. Adding to another's earnings did not reduce a subject's own earnings, and subtracting from another's earnings did not increase a subject's own earnings. The only cost of initiating an exchange was the "opportunity cost" of not exchanging with an alternative partner.

The amount of money an actor could add or subtract on any single opportunity, for any particular partner, was fixed and determined by the actor's structural position in the network. Actors could vary the value they produced for a partner only by varying the frequency with which they performed actions of fixed value.

Money is resistant to diminishing value (Assumption 2b) over the ranges typically offered in experimental sessions; therefore, I assume that the value of the outcome (and power) is stable during the course of the experiment. Recruitment announcements guaranteed subjects a minimum of $6 for participation and the opportunity to earn up to $20. Point values were set so that subjects earned, on the average, about $15 in a two-hour experimental session.

Exchange Structures

In all experiments, subjects participated in negatively-connected exchange networks composed of four actors (see Figure 1). Each actor could potentially exchange with two of the three other actors in the network; thus, each actor had access to two alternative exchange partners in the network. Negative connections were created by making the choice of partners mutually exclusive, so that exchange with one partner precluded exchange with another partner for that opportunity (Cook and Emerson 1978). Thus, the more frequently an actor exchanged with one partner, the less frequently she exchanged with the other.

Experimental instructions informed subjects that they were participating in a group of four persons, that each person would be able to interact with two of the other three persons, and that each person could act toward only one partner on each exchange opportunity. In reality, however, some of the actors in many of the experiments were computer-simulated actors, programmed to behave in specific ways. I used three different combinations of real and simulated actors for different experimental purposes. The early experiments in the program studied networks comprised of two real subjects (*A* and *B*) and two computer-simulated actors (see Figure 1a). The relation between the two subjects was the focus; the simulated actors functioned as alternative partners who responded with controlled strategies. In later experiments, I studied networks of four real actors, primarily to eliminate any possible effects of the simulated actors' programmed strategies on subjects' behavior (Figure 1b). These networks typically created two structurally-equivalent relations (*A—B* and *C—D*) which were the focus of the analysis. And, in some experiments that were designed to study the effects of an exchange partner's strategies on an actor's response to those strategies, I studied networks in which one real subject participated with two computer-simulated actors and a hypothetical fourth actor (Figure 1c).

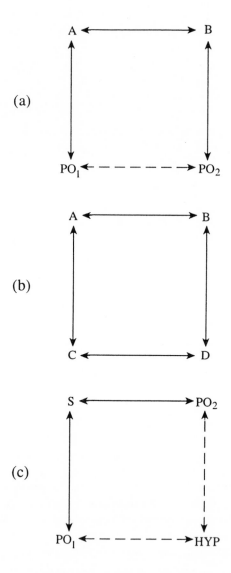

Note: The letters *A, B, C, D,* and *S* (for subject) indicate positions occupied by real actors. The *POs* in (a) and (c) are computer-simulated actors ("Programmed Others"), and *HYP* in (c) is a hypothetical actor. The dotted lines between the *POs* in (a) and between the *POs* and *HYP* in (c) indicate that those relations were implied rather than real.

Figure 1. Three Examples of Experimental Exchange Networks

Exchange Process

To meet the condition that exchanges are reciprocal and nonnegotiated, subjects individually performed rewarding or punishing actions, directed at specific partners, without negotiation and without knowledge of either partner's intentions. On each exchange opportunity, all actors in the network simultaneously chose (1) which partner they would act toward on that opportunity and (2) an action toward that partner (to add or subtract a fixed number of points). These two choices gave the subject a total of four behavioral options on each opportunity: reward partner *i*, punish partner *i*, reward partner *j*, or punish partner *j*. Because each side of an exchange was individually performed, subjects potentially could gain (or lose) money from one, both, or neither of their partners on each opportunity.

Subjects exchanged with the same two partners over a large number of exchange opportunities, typically 200-250. This number allowed exchange relations to develop gradually, based on the behavioral assumption (1.2) that actors learn from the consequences of trying different patterns of exchange with alternative partners. It provided sufficient time for exchange patterns to stabilize and met the scope condition of recurring exchanges between the same sets of actors.

Control Variables: Information

Subjects' information about their exchange partners, the exchange structure, the exchange process, and exchange outcomes was held constant. In all experiments, subjects were seated in isolated rooms, interacted with each other through computers, and did not meet before or after the experiment.

To allow for strategic power use (as well as the structurally-induced use of power that Emerson proposed), I provided subjects with information about the values of exchange controlled by both actors in each of their immediate exchange relations. Subjects did not know the dependencies of the fourth actor in the network, however, and consequently did not know how available that actor was as an alternative.

Subjects' information about the exchange process was consistent with the level of information that actors engaged in reciprocal exchanges typically would have. After all actors chose partners and behaviors on an exchange opportunity, subjects learned of their partners' choices; that is, whether each partner added *n* points to their earnings, subtracted *n* points from their earnings, or did not act toward them. Not acting toward one partner implied acting toward the other, but subjects were not told what action their partner took toward the other.

The experiment provided subjects with both per trial and cumulative information about their own outcomes from the exchange but with no information about their partners' outcomes. Subjects knew the consequences of their own actions for their partners, of course, but they did not know the net amount that their partners received from both of their partners on an exchange opportunity, nor did they know their partners' cumulative outcomes.

The Manipulation of Power

Theoretically, power in negatively-connected networks is a function of the *value* of exchange and *alternatives* to the exchange. Experimentally, I held constant the shape of the network and the number of alternative partners for each actor and manipulated power by varying the relative value of exchange with alternative partners. Within the *A—B* exchange relation, I operationally define *B*'s dependence on *A*, and *A*'s power over *B*, as equal to the value that *B* can receive from *A*'s behavior divided by the total value that *B* can receive from both of *B*'s potential partners in the network.

I manipulated the four dimensions of structural power—average reward power, average punishment power, reward power imbalance, and punishment power imbalance—by varying the number of points actors could add to each other's earnings (reward power) or subtract from each other's earnings (punishment power) on each exchange opportunity. On any single opportunity, each actor in the network could potentially gain a total of 10 points from both partners, and each actor could potentially lose a total of 10 points from both partners. Thus, a subject's net outcome on any single exchange opportunity could range from +10 points (if both partners added to her points) to -10 points (if both partners subtracted from her points). An actor's power over an exchange partner on either base (and the partner's dependence on the actor) consisted of the proportion of this total that the actor controlled. The imbalance on reward or punishment power in any dyadic exchange relation equaled the difference between the individual power-dependencies of the two actors on that base, and average power equaled the average of their power-dependencies. Actors' dependencies on their two partners always summed to 1.0, and their dependence on any single partner varied from 0 to 1.0. Similarly, power imbalance and average power ranged from 0 to 1.0.

Measures of the Dependent Variables

In all but the single-subject experiments, the unit of analysis is the exchange relation and the behavioral interaction in the relation is the focus of inquiry. Patterns of behavioral interaction provide the data for measures of both exchange *process* (particularly power strategies) and exchange *outcomes* (the frequency and asymmetry of exchange).

Measures of *exchange process* examine the dynamic contingencies between the two actors' behaviors; that is, how *A*'s behavior at time t_1 affects *B*'s behavior at time t_2 and vice versa. These contingencies are computed by examining the relation between actors' behaviors at lagged intervals across the entire period of exchange (see Molm 1990). In the later experiments that studied the effects of an exchange partner's strategy as an independent variable on a subject's response to that strategy, I manipulated these contingencies by varying the conditional probabilities with which computer-simulated actors responded to the behaviors of real subjects.

Measures of *exchange outcome* assess the overall frequencies of actors' behaviors toward one another, computed for the entire exchange period or for blocks of opportunities within it. The frequencies are divided by the number of exchange opportunities, thus standardizing the measures so that they vary from 0 to 1. For each base of power, I examined the average frequency of exchange on that base (i.e., the average frequency with which the two actors rewarded each other or punished each other) and the asymmetry of their exchange frequencies (i.e., the difference between the measures for the two actors). The asymmetry of reward exchange is my measure of *power use*. I measure power use by differences in frequency rather than value because actors in these experiments exchanged outcomes of fixed value. Higher power use, in *A*'s favor, indicates that *A* received the benefits of *B*'s exchange at lower opportunity cost.

In experiments in which a single real subject interacts with computer- simulated partners, the unit of analysis is the individual subject and the dependent variables are measures of the subject's behavioral or evaluative reactions to the strategies of the programmed partner(s). At the end of each experiment, subjects evaluated their partners' behaviors toward them on a series of semantic differential scales shown on their computer screens. These scales were designed to measure perceptions of justice and affective evaluations. I created multi-item scales from these responses with the aid of principal components analysis and reliability tests.

TESTING POWER-DEPENDENCE PREDICTIONS FOR COERCIVE POWER

In the first phase of the research, five experiments tested whether the central theorems of power-dependence theory hold for punishment power as well as reward power. I analyzed the predicted effects of coercive power (average power and power imbalance) on the average frequency and asymmetry of reward exchange and also examined the contingency of actors' rewards and punishments toward each other.

All of the experiments in this phase of the research studied exchange networks consisting of two real subjects and two computer-simulated actors (see Figure 1a). The focus of the research was the exchange between the two real subjects, *A* and *B*. Both the programmed strategies of the simulated actors and their structural relations to the real subjects were held constant in these experiments.[10]

Analyses of Exchange Frequency and Asymmetry

The first experiment (Molm 1988) tested the prediction that equivalent structures of imbalanced reward power and imbalanced punishment power produce equivalent power use. The experiment independently manipulated the balance/

imbalance of power on each base (holding average power constant at 0.6), in networks in which actors had both forms of power, and examined whether power imbalance had comparable effects for the two bases.

The results supported Theorem 3 for reward power but not for punishment power. As numerous studies had found previously, reward asymmetry increased with reward power imbalance (in favor of the more powerful actor), reward frequency decreased with reward power imbalance, and these effects increased over time. But punishment power imbalance had no effects whatsoever on either the asymmetry or the frequency of reward exchange. The frequency with which actors used punishment was very low (less than 0.05), it was unrelated to structural power or power position, and it declined significantly over time.

A second experiment (Molm 1989a) found that increasing the imbalance of punishment power from 0.4 to 0.8 increased the frequency of its use somewhat but not its asymmetry; that is, both actors were equally likely to punish each other even when punishment power was highly imbalanced. The absolute frequency of punishment was still low, and it declined over time. As in the first experiment, punishment power imbalance had no effect on reward exchange.

A third experiment (Molm 1990) tested the effects of variations in average punishment power (0.3 and 0.9) on the frequency of reward exchange. By Theorem 2, the frequency of reward exchange in a relation should increase with either average reward power or average punishment power. Other theories, such as conflict spiral theory (e.g., Deutsch 1973) and deterrence theory (e.g., Morgan 1977; Schelling 1960; Tedeschi, Schlenker, and Bonoma 1973), focus on the effects of average punishment power on *punishment* frequency. As Lawler's (1986) explication of these two theories illustrates, they make opposite predictions: conflict spiral theory predicts a positive relation between average punishment power and punishment frequency, and deterrence theory a negative relation. I found support for none of these predictions. Like punishment power imbalance, average punishment power had no effects on the frequency or asymmetry of either rewarding or punishing. As in the first two experiments, actors rarely used punishment, and its frequency declined over time.

Finally, the fourth and fifth experiments (Molm 1989b, 1990) examined whether the two bases of power interact in ways that are not predicted by a simple extension of power-dependence theory and that were obscured in the first three experiments. Would punishment power have stronger effects under different conditions of reward dependence, that is, when average reward power was higher or lower or reward power more or less imbalanced? These experiments found the first significant effects of punishment power. Average punishment power did affect the use of punishment, but the direction of these effects depended on the structure of reward power: actors were more likely to use low rather than high magnitudes of punishment power when their mutual dependence was greater (i.e., either average reward power was high, or power imbalances on the two bases favored different actors so that each could inflict more than trivial damage on the other). Under some conditions, average punishment power increased reward frequency, and a punishment power advantage that opposed a reward power advantage increased average reward exchange and decreased the other actor's exchange advantage—results that are consistent with power-dependence predictions. As before, however, the use of punishment was low in all conditions, and its frequency declined over time.

Analyses of Exchange Process

Using a combined data set from all five experiments, I conducted a series of logistic regression analyses to estimate the contingency of each actor's behavior at time t on the other actor's behavior at time t -1 (Molm 1990). These analyses assessed the extent to which each actor's use of punishment was influenced by the other actor's immediately prior punishment (retaliatory punishment) or immediately prior nonexchange (coercive punishment).

The results showed, first, that actors were much more likely to respond reciprocally to a partner's rewards or punishments than to use power, which required actors to respond nonreciprocally (with punishment or reward withholding). Although either punishment or nonexchange by the partner at time t -1 increased the probability of the actor's punishment at time t, retaliatory punishment was far more common than coercive punishment. Second, the structure of power had no effect on the likelihood that an actor would use retaliatory punishment, but the use of coercion increased with an actor's reward dependence, and, to a lesser extent, with a punishment power advantage. In other words, actors were more likely to use coercive punishment when they had more to gain from the other's increased reward exchange (reward dependence is high) and less to lose from the other's retaliatory punishment (punishment power is imbalanced in their favor). Reward withholding was affected solely by reward power imbalance; as reward power imbalance increased, advantaged actors were less likely to reciprocate rewarding and more likely to withhold rewards. Third, net of structure, both retaliatory and coercive punishment affected reward frequency and reward asymmetry, while reward strategies had no effect. Reward frequency increased with the use of both retaliatory and coercive punishment, and reward asymmetry increased in favor of the actor who used these strategies more frequently.

Implications

Four main conclusions emerged from these five experiments: First, across a wide range of variations in dimensions of both reward power and punishment power, the effects of punishment power failed to equal or even approximate the effects of comparable dimensions of reward power. Most analyses found no effects of punishment power on either the frequency or the asymmetry of reward exchange.

Second, despite the persistent weakness of punishment power, the cumulative pattern of findings (including some weak interactions not discussed in this brief summary) suggests that the effect of punishment power on reward exchange—when one occurs—is positive. Reward exchange either increased with average punishment power or was unaffected by it. Imbalanced punishment power either benefited the actor advantaged on punishment power or had no effect on reward exchange. And the contingent use of punishment, to both coerce and retaliate, increased the partner's reward frequency. The experiments provide no support for the negative effects of punishment power on reward exchange that Homans, Blau, and other theorists proposed; that is, punishment power never decreased the frequency of reward exchange or reduced actors' benefits from exchange.

Third, the findings fail to support the predictions of some theories (e.g., Deutsch 1973) that the use of punishment leads to escalating cycles of conflict. In all of the experiments, the use of punishment was extremely low (on the average, actors used punishment on fewer than 0.05 of all exchange opportunities), and in all experiments, punishment declined rather than increased over time.

Fourth, as with reward exchange, variations in the use of punishment nevertheless fit a pattern that is consistent with the central assumptions of social exchange theory: punishment was used more frequently under structural conditions that increased the gains of using punishment power and decreased the losses. The frequency of punishment increased with punishment power imbalance and with reward power disadvantage, and actors used more potent (and riskier) magnitudes of punishment when mutual dependence was lower or relational conflict was greater. Similarly, the use of coercive punishment increased with reward dependence and with punishment power advantage. The one exception to this pattern was the consistent tendency for actors in unequal positions of punishment power to use punishment equally often.

In summary, these experiments found that under the conditions assumed in this research program, reward power and punishment power have effects on reward exchange that are similar in direction but vastly different in magnitude. These results suggested that power-dependence theory could not be extended, without modification, to predict the effects of coercive power.

New Directions

With these conclusions, the focus of the research program shifted to the task of explaining these differences and analyzing the implications for exchange theory. In the next phase of the research, I developed and tested a theory of coercive power that rested on three central arguments.

First, the use of coercive power is not structurally induced. Unlike reward power, the use of coercion is not inherent in a structural advantage on punishment power; that is, the power to punish does not lead automatically to its use. *Second*, the strategic use of coercion is constrained by risk. This risk is directly related to the structural conditions that provide the incentive to use coercion—reward dependence and reward power disadvantage—and to the loss aversion of actors choosing among alternatives with uncertain outcomes. Justice norms support and maintain the effects of dependence by increasing the likelihood that disadvantaged actors' use of coercive strategies will be perceived as unfair and sanctioned by their advantaged partners. *Third,* the weak effects of coercive power result from the low use of coercion by actors, not from the effects of coercion on their targets. When used consistently, the effects of coercive power on exchange are at least as strong as those of reward power.

In the next sections I develop these arguments and summarize the research testing them.

THE STRUCTURAL DETERMINATION OF POWER USE

Emerson argued that as long as actors follow basic behavioral principles, the incentive to use power is inherent in the structural potential for power. The same structural conditions that provide actors with the *capacity* to use power also provide the *incentive* to use power, whether knowingly or not. Power use occurs as an unintended byproduct of advantaged actors seeking exchange with alternative partners in structurally imbalanced relations.

This effect of structural power rests on the behavioral principles of Assumptions 1.1 and 1.2. When actors in relations of imbalanced reward power follow these principles, their interaction will produce the distribution of exchange predicted by power-dependence theory: an asymmetry in exchange that favors the less dependent actors. Advantaged actors will "use power" more than disadvantaged actors; that is, they will more frequently initiate exchange with alternative partners, thus inadvertently withholding rewards from their disadvantaged partners. Their disadvantaged partners may temporarily reduce their rewarding in response, but eventually they will accept a pattern of intermittent reciprocity for their exchange as long as that pattern is more rewarding than alternatives.

The effects of coerced exchanges rest on the same principles as reward exchanges; that is, the targets of coercion give rewards to their coercers in exchange for the cessation or avoidance of punishment. Thus, an imbalance in punishment power should lead the disadvantaged actor to give increased rewards to the advantaged partner to prevent punishment. But it does not induce the *use* of punishment power; that is, it does not induce the advantaged partner to use punishment if the disadvantaged actor fails to provide rewards. And without use, any effects of potential punishment power are unlikely to be maintained.

In short, unless we bring in some notion akin to conflict spiral's concept of "temptation" (e.g., Deutsch 1973; Lawler 1986), the incentive to punish is not inherent in a structural advantage on punishment power or any other condition of structural power. Structure can provide the motivation to coerce by producing unequal or insufficient rewards from exchange, but it does not lead automatically to coercive behaviors. If punishment power is used, it must be used strategically. *In this respect, reward power and punishment power are fundamentally different.* The use of coercion does not follow logically from the capacity to coerce, as an unintended byproduct of actors following the behavioral assumptions that underly exchange theory. *If used,* the effects of punishment power imbalance should be comparable to those of reward power imbalance, because both determine an actor's capacity to minimize his or her own costs while imposing cost on another. But the principle that a structural power advantage provides its own incentive for power use holds only for reward power.

This is a logical argument and requires no empirical test. However, an important inference of this logic can be tested: *if* the use of both bases of power were structurally induced by a power advantage on that base, then their effects should be equivalent.

An Experimental Test

Testing this prediction requires creating exchange networks that will structurally induce the use of punishment. No such networks exist naturally, but they can be created artificially by designing networks that meet two assumptions. First, *the exchange advantage of using punishment power must be avoiding harm, not increasing rewards.* A network based solely on punishment power meets that condition. Eliminating reward power changes the function of punishment power from coercing gains to minimizing losses. As a result, a punishment power advantage provides both the incentive and the capacity to use power, just as a reward power advantage does. I compared networks in which actors had only punishment power with networks in which actors had only reward power.

Second, to create comparable conditions in these "single-base" networks, *all actors were required to initiate exchange with one partner on each opportunity.* That condition is implicit in reward-based networks; that is, actors who are motivated to increase rewards should always take the opportunity to initiate exchange when rewards are controlled by other actors and can be obtained solely through exchange. But in networks based exclusively on punishment power, exchange relations are unlikely to form without that requirement.

The experimental design compared networks of four real actors who had only reward power or only punishment power over their partners with networks in which they had both (Molm 1993). Within each of these networks, power in the focal relations was either balanced or imbalanced. All actors were required to initiate some action toward one of their partners on each exchange opportunity.

In the networks based solely on reward power or punishment power, a power advantage on either base should induce power use. Just as actors in reward-based networks are expected to increase exchange with the partners who can reward them the most and to decrease exchange with the partners who can reward them the least, actors in punishment-based networks should increase exchange with the partners who can harm them the least and decrease exchange with the partners who can harm them the most. These behaviors will inadvertently result in the power-advantaged actors "using power" in their relations with their disadvantaged partners by imposing greater costs on them through reward withholding or punishment. The result, in both networks, should be asymmetrical exchange that favors the power-advantaged actors; that is, these actors should receive more rewards or fewer punishments than they administer.

The results of the experiment showed strikingly similar effects for the two bases of power in the single-base networks. In both the reward-based and punishment-based networks, power imbalance produced comparable effects, decreasing the average frequency of exchange and producing asymmetrical exchange that favored the advantaged actor. Power base had no effects on exchange. Within conditions of power balance or imbalance, the two bases of power produced frequencies and distributions of exchange that were virtual mirror images of each other (i.e., both produced exchange asymmetries of 0.46, in favor of the power-advantaged actor). In the comparison networks in which actors could both reward and punish each other, the findings of the earlier experiments were replicated: punishment power was rarely used, and only reward power imbalance affected the distribution of exchange.

Implications

As this experiment shows, when structural conditions are created that induce the use of both bases of power, they have virtually identical effects on behavior. Actors who are faced with the task of simply distributing rewards or punishments among alternative partners follow the same principles in doing so: they provide more frequent benefits (rewards or the withholding of punishment) to those on whom they are most dependent for obtaining rewards or avoiding punishment.

Actors in the punishment networks were not using coercion to obtain rewards, however; they were simply minimizing losses. In the networks of interest in this research program—networks in which all actors have the power to both reward and punish, and punishment power can be used to coerce rewards— the incentive to use coercion must come instead from the motivation to increase rewards. It is the structure of *reward* power, not punishment power, that motivates actors to use coercion. Reward power does not affect the use of coercion directly, however, but indirectly through its effect on reward exchange. When valued rewards are received only infrequently or unequally, the motivation to use coercion to increase the partner's rewarding should increase. But that requires the use of contingent influence strategies.

THE STRATEGIC USE OF COERCIVE POWER

Two important differences distinguish strategic power use from the structurally-induced use of power with which Emerson was concerned. First, actors do not merely respond to the consequences of their own behaviors; they *create* contingencies that *produce* consequences for *other* actors' behaviors (Thibaut and Kelley 1959). As the term "strategy" suggests, such actions are typically purposive. With few exceptions, actors who use power strategically impose cost knowingly.[11] Second, strategic power use is based not on the expectation of reciprocity that underlies Assumption 1.1 and the inadvertent use of structurally-induced reward power but on an application of the behavioral principles in Assumption 1.2. Rather than giving rewards to obtain benefits in return, actors impose cost (by administering punishment or withdrawing rewards) on partners who fail to provide sufficient benefits.

Assumption 1.3.   Actors initiate strategic power use by imposing cost on other actors who fail to provide rewards in exchange.

Just as A's initiations of reward exchange with B should increase with A's dependence on B (Theorem 1), A's initiations of coercive power strategies against B should increase with structural conditions that increase the value of B's rewards for A and decrease the probability that B provides those rewards:

Theorem 4.   The probability that A initiates a coercive power strategy to influence B increases with A's dependence on B and with A's reward power disadvantage.

Once initiated, strategies—like other patterns of exchange behavior— should increase or decrease according to their consequences (Assumption 1.2). Coercive strategies that increase rewards should persist; those that provoke retaliation (i.e., decreased rewards or increased punishment) should decline. In addition, to the extent that coercion is purposive, *anticipated* gains and losses can also encourage or deter the use of coercion. Consequently, structural conditions that increase the magnitude or probability of potential gain, and/or decrease the magnitude or probability of potential loss, should increase actors' use of coercive power strategies. Conversely, conditions that increase loss should decrease coercion.

Dependence and Risk

Because actors who use power impose cost on another, power use is always risky. Rather than improving an actor's outcomes by increasing the partner's rewards or reducing the actor's costs, the use of power might instead produce the opposite result by provoking retaliation. Risk increases with dependence; the greater A's dependence on B, the greater the loss that B can inflict if B retaliates against A's power use (by withdrawing rewards or administering punishment). In relations in which actors have both reward and punishment power over each other, the potential loss from using power increases with dependence on the partner for *either* obtaining rewards or avoiding punishments. The magnitude of potential loss from *B*'s retaliation increases with *A*'s absolute dependence on *B*, and the probability of *B*'s retaliation increases with *A*'s relative dependence (or *B*'s power advantage).

These risks are unlikely to deter the use of reward power, however, because a reward power advantage reduces risk as it increases incentive. Disadvantaged actors are less likely to retaliate, and if they do, the costs they can impose are lower. Consequently, whether reward power is used strategically or induced structurally, its use should increase with power advantage.

In contrast, the structural conditions that provide the incentive to use coercion— high reward dependence and reward power disadvantage (Theorem 4)—also increase the potential loss from power use. Even if actors are advantaged on punishment power (thus reducing their potential loss from the partner's punitive retaliation), they face potential loss from the partner's reward withdrawal. As Blalock (1987, p. 13) observed, "substantial degrees of dependence will lead to both a high level of motivation to employ negative sanctions to the other party *and* high potential costs of such actions ... where exchanges of positively valued goods accompany the punitive actions, the more dependent party is especially vulnerable to a cutoff of the exchange relation altogether."

Consequently, under most conditions, actors are unlikely to use coercion with any persistance. Not only are losses more likely, but loss aversion doubles their weight. The result is a bias toward the status quo (Kahneman and Tversky 1984) and a decrease in the use of punishment over time, as all of the earlier experiments found.

Actors are more likely to use coercion under some conditions than others, however. The actual loss that *A* can impose on *B* through reward withdrawal, in retaliation for *B*'s use of coercion, depends not only on the structure of dependence but on the extent to which *A* has already withdrawn rewards. Because gains and losses are relative to the status quo (Assumption 2a), potential gains increase and potential losses decrease as *B*'s current outcomes from exchange with *A* decline. The conflict between potential gain and potential loss is highest when the partner's reward frequency is moderate (as it was in most of the earlier experiments) and declines as it either increases or decreases.

Let us assume that the range of *A*'s potential rewards to *B*, standardized by *A*'s opportunities to reward *B*, varies from 0 (*A* never rewards *B* on available opportunities) to 1 (*A* always rewards *B*). If p equals the proportion of opportunities on which *A* currently rewards *B*, then, holding *B*'s dependence on *A* constant, *B*'s maximum potential for gain is $1-p$, and *B*'s maximum potential for loss from *A*'s reward withdrawal is p. When *A*'s reward frequency is high, *B*'s potential losses clearly exceed *B*'s potential gains. But as p approaches zero, the amount that *B* can lose from *A*'s reward withdrawal becomes trivial, even if *B*'s structural dependence is high and the probability of *A*'s retaliation is high. (*B*'s potential losses from *A*'s punitive retaliation, which vary with the structure of punishment power, still remain, of course.) Because loss aversion roughly doubles the value of any potential loss, however, A's reward frequency must be quite low before that occurs.

Theorem 5. The positive effect of reward power disadvantage on both the potential gains and the potential losses from using coercive power will constrain the use of coercion unless the proportion p of opportunities on which the partner rewards the actor, multiplied by the coefficient for loss aversion c, is less than $1 - p$: $pc < (1 - p)$. As p declines below that level, the use of coercive power should increase.

If c equals 2 to 2.5, as a number of analyses suggest, then Theorem 5 implies that B is unlikely to use coercion against A unless the frequency with which A rewards B falls below 0.3 or so (i.e., if c equals 2, then p equals 0.33; if c equals 2.5, then p equals 0.29). Below that level, the probability of coercion should increase as A's reward frequency decreases. Such low levels of rewarding are likely to occur, however, only when A's reward dependence is very low—too low to support A's continued exchange with B, even if that exchange is asymmetrical (Michaels and Wiggins 1976). In previous experiments in this program, reward frequencies of that level were rare.

In contrast, the predicted effects of punishment power advantage (which affects only potential loss) on the use of coercion are straightforward:

Theorem 6. The use of coercive power increases with punishment power advantage.

Experimental Tests

Two experiments tested these predictions (Molm 1993). The first experiment examined the predicted effects of reward power and punishment power (Theorems 5 and 6) on the use of coercion. Four real actors were assigned to positions in a network structure that provided two of the actors with both the incentive (reward power disadvantage) and the capacity (punishment power advantage) to use coercion against their partners. The 2×2 factorial design manipulated the potential losses they incurred if they did so: (a) the potential loss from their partner's retaliatory punishment (manipulated by varying the partner's punishment power, which was either 0.4 or 0) and (b) the potential loss from their partner's withdrawal of valued rewards (manipulated by varying the partner's reward dependence [0 or 0.4], which in turn affected the partner's reward frequency).

Of these two sources of loss, the potential loss of the partner's reward exchange had by far the stronger effect on actors' use of punishment. With this source of loss minimized, however, the effect of the partner's capacity to retaliate punitively increased over time. Reducing both sources of loss increased punishment fourfold, from 0.06 to 0.23.

The second experiment tested the effects of structural variations in reward power and punishment power when both sources of potential loss are reduced. Actors with either reward power or punishment power (with one base high and the other low in power or both equal) were placed in competition for the reward exchange of a subject who had no capacity to retaliate punitively. With risk reduced and the frequency of coercive punishment increased (to 0.27), the subject's choice of exchange partners depended on the relative strength of their power, not its base. Strong coercive power was as effective as strong reward power.

Implications

These results support Theorem 5 and suggest that the structure of reward power has a much stronger effect on the use of coercion than does the structure of punishment power. As Theorem 5 predicts, the conflict between gaining a more rewarding relation and losing the current level of rewards tends to constrain coercion until the partner's reward frequency is low enough that potential losses from the partner's reward withdrawal—even when weighted by loss aversion—are lower than potential gains.

The irony of this analysis is that it implies that coercion should increase to effective levels only when the structural basis for mutually beneficial reward exchange is weak or absent. As long as the partner's reward frequency is greater than 0.3, the potential losses from using coercion are likely to outweigh the potential gains. At very low reward frequencies, this relation reverses. Low reward frequencies typically imply low reward dependence, however, and the lower the partner's reward dependence, the more difficult it should be to coerce the partner into reward exchange.

Consequently, coercion is most likely to be used in purely coercive structures, in which there is no relation of mutual reward exchange to risk losing (see Willer 1987). Coercion can be effective even under these conditions, but the coercer must rely solely on the threat of punishment to maintain the other's rewarding, rather than using coercion as a tool for establishing a rate of reward exchange that is more favorable to the coercer. Punishment should be far more effective when combined with rewards and used in a structure that supports mutually beneficial exchange (Molm 1994a; Van Houten 1983).

Justice Norms and Risk

The classical exchange theorists (Blau 1964; Homans 1974; Thibaut and Kelley 1959) proposed that actors evaluate their outcomes not only in relation to the status quo but in relation to expectations derived from past experiences or social norms. When expectations are based on *social* comparisons of some kind, questions of fairness or justice arise. Outcomes that depart from standards of "fair exchange" produce emotional reactions, such as anger, and these emotions can lead to behavioral efforts to restore justice. Two experiments in the program examined how justice evaluations and behavioral reactions to injustice affect the costs of using coercion.

The assumption that perceptions of injustice will lead to behavioral efforts to restore justice is based not on principles of learning, however, but on principles of cognitive consistency and social comparison. Thus, despite the exchange roots of justice theories, their predictions are derived from a rather different set of assumptions.

Justice Assumptions

Perceptions of justice, like evaluations of outcomes, are referent dependent. Actors make judgments about the fairness of distributions of outcomes by comparing their actual outcomes to some justice standard; that is, what they think they ought to receive (e.g., Jasso 1980; Markovsky 1985). The reference standard for justice evaluations is some function that compares the actor's own outcomes with those of another actor, group, or referential structure (Hegtvedt and Markovsky 1995). Departures from this standard should create feelings of injustice, which in turn should lead to behavioral efforts to restore justice.

I make two assumptions about the justice standard in relations of reciprocal exchange. First, I assume that the basic standard of justice is derived from the scope condition that defines social exchange; "benefits obtained ... are contingent on benefits provided" (Emerson 1981, p. 32). When transformed from a description of social exchange to a prescription for fair exchange, this "norm of reciprocity" (Gouldner 1960) specifies that outcomes received should be contingent on, and functionally equivalent to, outcomes given (i.e., good should be repaid by good and harm by harm) (Molm, Quist, and Wiseley 1993).

Second, I assume that actors' structural positions affect their expectations of the process and outcomes of exchange (Berger, Cohen, and Zelditch 1972; Cook 1975). In imbalanced relations, those expectations can shift the standard of justice away from reciprocity, in a direction that is congruent with actors' relative positions of power. Departures from reciprocal exchange may still be judged unfair (e.g., Cook and Hegtvedt 1986; Stolte 1983, 1987), but they are more likely to be judged unfair when they favor actors who are disadvantaged on power than when they favor those who are advantaged.

Assumption 5: The Justice Standard. In balanced relations of direct exchange, the standard of fair exchange is defined by reciprocity, that is., the contingency of outcomes received on outcomes given. As power imbalance increases, expectations attached to positions of power shift this standard in a direction that is consistent with actors' positions of relative power.

Assumption 6: The Comparison Process. As discrepancies between the standard of fair exchange and actual exchange increase, those who receive fewer rewards or more punishments will feel emotional distress.

Assumption 7: The Behavioral Reaction. Actors seek to reduce the emotional distress produced by departures from fair exchange by reducing their rewards or increasing their punishments toward the partner.[13]

Justice and Power Use

When actors use power, they violate the standard of reciprocity in exchange. Actors who withhold rewards reciprocate the other's reward exchange only part of the time; actors who punish coercively respond to the partner's failure to reward with punishment. Consequently, when *A* uses power against *B*, that behavior should provoke feelings of injustice in *B*, and *B* should respond with behavioral efforts to restore justice (by retaliating with reward withdrawal or punishment). Thus justice norms, like dependence, should increase the potential costs of using power.[14]

These effects should be stronger for coercive power, however, and they should vary with the structure of power. Objectively, the use of either reward power or coercive power involves equal departures from reciprocity if the probabilities of contingent punishment or reward withholding are equal and the values of the rewards and punishments are equal. Because of loss aversion, however, actors should tend to perceive coercion as more nonreciprocal than reward withholding. Therefore:

Theorem 7. The perceived injustice of exchange partners' power use is greater for coercive power than for reward power.

And, because power use is more consistent with the expectations attached to power-advantaged positions than to disadvantaged positions:

Theorem 8. The perceived injustice of exchange partners' coercion increases with the power disadvantage of the coercer.

Experimental Tests

Two experiments tested Theorems 7 and 8. In the first experiment, computer-simulated actors employed either tit-for-tat (reciprocal), coercive, or reward-withholding strategies in networks of balanced power relations (Molm, Quist, and Wiseley 1993). In balanced networks, the standard of fair exchange should be reciprocity, and perceptions of injustice should increase with departures from reciprocity. As predicted, subjects judged both power strategies (coercion and reward withholding) as more unfair than tit-for-tat strategies, but they judged coercion as far more unfair than reward withholding.[15]

A second experiment examined how the structure of power influences actors' evaluations of their partners' power strategies (Molm, Quist, and Wiseley 1994). Computer-simulated actors again used programmed power strategies, this time against subjects who were advantaged, disadvantaged, or equal to them on each of the two bases of power. As predicted by Theorem 8, the structure of power influenced the perceived injustice of coercive strategies. Coercive strategies were perceived as most unjust when used by simulated actors who were disadvantaged on reward power against their advantaged partners, as least unjust when used by advantaged actors against their disadvantaged partners, and as intermediate in justice when used by actors in balanced relations. In contrast, the structure of punishment power had no effect on evaluations of justice, suggesting that only reward power affects justice standards.

In both experiments, perceptions of injustice increased actors' tendencies to respond to a partner's coercion with resistance, that is, with lower-than-expected levels of rewarding. The more unjust an actor viewed a partner's power strategy, the more likely she was to resist rather than comply. Perceptions of injustice did not affect punitive retaliation, however. Regardless of their perceptions, power-advantaged actors were more likely to both resist and retaliate a disadvantaged partner's use of coercion. Over time, the actual consequences of retaliation tended to overcome emotional reactions; when retaliation was costly, it rapidly declined.

Implications

These analyses show that justice norms support and maintain the effects of reward dependence on the use of coercion. Actors who are disadvantaged on reward power face greater risk of reward loss *and* greater risk of normative censure if they use coercion against their advantaged partners. The latter increases the former. But even without the mediating effects of injustice, disadvantaged actors who use coercion are more likely to be met with both resistance and punitive retaliation from their advantaged partners. These responses tend to decline over time, but their initial occurrence is likely to deter the use of coercion by loss averse actors. In short, coercion *is* costly, at least in the short run, and these costs are greatest for the very actors who have the strongest incentive to use it: actors who are disadvantaged on reward power and on the losing end of the exchange.

THE EFFECTS OF COERCION ON SOCIAL EXCHANGE

The final series of experiments tested the argument that it is the low use of coercive power, not the ineffectiveness of coercion per se, that accounts for its weak effects on social exchange. Results of the earlier experiments indicated that both bases of power have equal effects when either is the sole source of power and that coercive power is more effective when the risk of retaliation is removed and its use increases. But how effective is coercion in the networks that are the focus of this research—networks in which all actors have both reward and punishment power and retaliation is possible? If actors in these networks used coercion more frequently and consistently, would their coercive power be more effective?

No new assumptions are required to predict the effects of coercive power use on social exchange; that is, actors should increase the frequency of their reward exchange either to obtain rewards or to avoid punishments. If anything, loss aversion should make coercion *more* effective than reward withholding; that is, the same aversion to loss that constrains the use of coercion should make it highly effective when it is used. Thus, we can predict:

Theorem 9. The effectiveness of coercion increases with the contingency of its use; that is, as the contingency of A's punishment on B's nonrewarding increases, the frequency of B's rewards to A will increase.

Theorem 10. The effects of coercive power (i.e., average punishment power and punishment power imbalance) on reward exchange increase with the contingent use of coercion.

These predictions stand in sharp contrast to the positions of many social scientists. The exchange theorists who excluded punishment from their analyses, as well as many theorists who explicitly study the use and effects of coercive power (e.g., Deutsch 1973; Lawler 1992; Pruitt 1981; Tedeschi, Schlenker, and Bonoma 1973), argue that it will be *less* effective the more it is used. Rather than increasing compliance, they propose that coercion inhibits cooperative behavior and leads to escalating conflict and social disruption. Emotional reactions (hostility, anger) and impression management concerns (saving face, appearing tough) lead actors to retaliate a partner's punishment, regardless of the cost, and retaliation leads to counter-retaliation. Consequently, when the power to punish is bilateral, increasing the strength and frequency of coercion should not increase compliance. Instead, coercion is likely to produce an escalation of conflict, decrease mutual reward exchange, and produce lower benefits for both actors.

Experimental Tests

I tested Theorems 9 and 10 in exchange relations in which reward power was imbalanced and the disadvantaged partner (a computer-simulated actor) used coercion to induce the advantaged partner (a real subject) to reward more frequently (Molm 1994a). I examined the effects of three different levels of coercion by the disadvantaged partner (i.e., the *PO* contingently punished the *S*'s prior nonrewarding with probabilities of 0.1, 0.5, or 0.9) in each of four different structures, varying on dimensions of punishment power. Punishment power was either balanced or imbalanced (the subject had equal or less power to retaliate), and average punishment power was either high (0.7) or low (0.3).

The results strongly supported the exchange predictions and refuted the conflict escalation predictions. As Theorem 9 predicts, the *PO*'s coercion had a strong and positive effect on subjects' rewarding. As the probability of coercion increased from 0.1 to 0.5 to 0.9, the mean frequency with which subjects rewarded the coercive *PO* increased from 0.20 to 0.36 to 0.63. Furthermore, compliance increased over time when the probability of contingent punishment was high, decreased when it was low and did not change when it was moderate.

As Theorem 10 predicts, the increased use of coercion also enhanced the effects of coercive power; interactions between the use of coercion and both average punishment power and punishment power imbalance were significant. The latter interaction was weaker, primarily because the unswerving toughness of the robot coercer tended to wipe out the effect of power imbalance at the highest level of coercion. Under these conditions, the capacity to retaliate with equal harm becomes irrelevant; that is, resisting a partner who continues to punish at very high levels, no matter what, is simply too costly. Real actors, unlike this *PO*, would be expected to vary the contingency of their punishment as the structure and their partners' behavior changed.

Not only was strong coercion more effective than weak coercion, but it produced no more retaliatory punishment. This experiment, like the earlier ones, suggests that some retaliation for coercion is common, but its frequency is low (0.06 in this experiment), it declines rapidly over time, and it is largely independent of structural power.

Even more surprising, strong and consistent coercion produced *less* negative affect than weak and inconsistent coercion. In these experiments, punishment for nonrewarding was combined with rewards for rewarding; thus, subjects who complied with coercion soon became engaged in mutually (albeit unequally) rewarding exchanges with the *PO*. Because consistent coercion was far more effective than sporadic coercion, its targets experienced more rewards than punishments from the *PO*. As other studies have found (e.g., Lawler and Yoon 1993; Patterson 1982), positive affect increases with the relative frequency of positive to negative experiences.

Three additional experiments, which tested the generalizability of these findings under further variations in coercive strategy and structure, suggest that these effects are quite robust. Coercion is highly effective as long as its contingency is maintained; reducing the association of punishment with undesirable responses reduces its effectiveness.

Implications

These experiments show that when coercion is applied consistently and contingently, it is a powerful means of obtaining greater benefits from an exchange partner, even when that partner is advantaged on reward power. By demonstrating that weak coercion actually *decreases* reward exchange (at least when used against a reward-advantaged partner) while strong coercion increases it, the results support the argument that it is the infrequent use of coercion, not the ineffectiveness of coercive power per se, that accounts for its weak effects in relations with both bases of power.

SUMMARY AND CONCLUSIONS

By proposing that power is an attribute of a relation derived from actors' mutual dependencies on each other for valued outcomes, social exchange theorists extended its scope and application in significant ways. They conceptualized power not as the province of particular actors with substantial resources or political clout but as a ubiquitous part of virtually all social relations.

While broadening the scope of power in some respects, however, exchange theorists narrowed it in another. Rather than extending the more traditional conception of power as coercive to include power derived from dependence on another for rewards, they omitted coercion from the theory and its analysis of power. Consequently, the study of reward-based power and punishment-based power became the subject of separate theories and separate research programs with few theorists examining the relation between the two (e.g., Bacharach and Lawler 1981; Willer and Anderson 1981).

The decade-long program described here sought to bring coercive power within the scope of exchange theory and, in the process, to understand the similarities and differences between the two bases of power. Empirically, the strongest and most consistent finding is that when actors are dependent on one another, both for obtaining rewards and avoiding punishment, the structure of reward power dominates interaction. Punishment and coercion are rarely used, even by actors who are disadvantaged in the reward exchange, and the structure of punishment power has almost no effects on the frequency or asymmetry of reward exchange.

The very different effects of the two bases of power stem from two important theoretical distinctions between them. First, the use of coercion is not structurally induced by punishment power advantage. Unlike reward power, the *capacity* to use punishment power does not provide the *incentive* to use it. That incentive comes primarily from the effects of reward power disadvantage, that is, high reward dependence on an actor who fails to provide those rewards in exchange. Second, because actors with the incentive to use coercion are dependent on another for rewards, the use of coercion is risky. Reward dependence provides the motivation to coerce, but it also increases the costs that a partner can impose in retaliation. Under these conditions, loss averse actors typically opt for the status quo, forgoing coercion and the gains it might bring to avoid potential loss. Justice norms reinforce these effects. Actors perceive coercive tactics as unjust, and as their sense of injustice increases, so does their resistance to coercion. These reactions are strongest when advantaged actors, who expect rewards rather than punishments from their more dependent partners, are the targets of coercion.

Consequently, actors are unlikely to use coercive power more than sporadically unless their partner's reward frequency is so low that the potential loss of those rewards is of little consequence. This condition describes purely coercive relations (in which each actor has only a single base of power), as well as structures in which mutual reward dependence is highly imbalanced but too low to support mutually beneficial reward exchange. In such structures, there is little or no relation of reward exchange to risk losing. If, instead, the joint reward dependence of actors is high enough to sustain an exchange relation (even an asymmetrical one), then the risk of losing the partner's rewards, even if those rewards are insufficient and unequal, is likely to be great enough to keep the use of coercion relatively low and ineffective. Thus, ironically, coercion is most likely to be used under the very conditions in which it is least likely to be effective.

As this discussion indicates, my analysis locates the causes of differences between reward and punishment power in conditions that affect their *use*, not their *effects*. In contrast to theorists who argue that punishment increases hostility and leads to escalating conflict, my results show that coercion is, instead, a highly effective means of securing rewards from an exchange partner. While loss aversion constrains the use of coercion, it enhances its effects. Still, coercive strategies influence behavior only gradually, and some initial retaliation is typical. For loss averse actors who are uncertain of the long-term benefits of coercive strategies, the immediate losses that typically accompany coercion are sufficient to suppress its further use.

What are the implications of this program for the objective of extending exchange theory to incorporate coercive power? As Heath (1976) proposed, no new principles are required to explain the *effects* of coercive exchange. Actors will increase the frequency of their reward exchange to obtain rewards or to avoid punishment. But the relation that Emerson proposed between the *structure* of power and its *use* holds only for reward power. The use of coercive power is not structurally induced by punishment power advantage as an inadvertent byproduct of actors' pursuit of their own self-interest. Coercion is a purposive power strategy in which actors knowingly and contingently impose cost on another to influence the other's behavior. Nevertheless, like other behaviors, it is influenced by its consequences, and the structure of dependence remains central to predicting those consequences.

Analyzing coercive power within the framework of social exchange relations also offers a different perspective on coercion. Many of the connotations that typically accompany analyses of coercive power (that it is used by the powerful to subjugate the weak, that it employs extreme sanctions, that it is morally repugnant) are derived from the study of very particular, highly visible forms of coercion, such as the use of military power by the state or the torture of captives by captors. As a form of power in social exchange relations—relations which also entail mutual reward dependence—coercion emerges as a far less malevolent force. It is primarily a tool of the disadvantaged, that is, of actors who lack the reward power to get what they want. Their use of coercion is not legitimated by a position of power advantage, and it is more likely to provoke retaliation. Consequently, it is not surprising that disadvantaged actors in relations of unequal reward power rarely "fight back," or that their use of coercive tactics—which tends to be sporadic and weak—is often described in terms that capture both the negative evaluation and the impotence of their efforts (e.g., "nagging"). Their use of coercion differs, both objectively and subjectively, from the use of institutionalized coercion by powerful actors.

By systematically comparing equivalent forms of reward and coercive power in exchange structures in which both are available, this research also helps to explain some of the puzzling inconsistencies and contradictions that are evident in much of the literature on coercive power. Various theorists have proposed that coercion is either extremely powerful or ineffectual and that its use is either deterred or encouraged by actors' power to coerce. My analysis shows that these seemingly incompatible conclusions are sometimes both true under different conditions. Another's capacity to punish does deter the use of coercion, but a far more important factor is the reward power in the relation. And while coercion is a potentially powerful means of influencing others, it is weak when used inconsistently or under structural conditions that fail to support it. Paradoxically, coercion is most likely to be used under the very conditions in which it is least effective, contributing to the perception that coercive tactics do not work in exchange relations.

Finally, studying coercion in exchange relations strips punishment and coercion of some of the affective evaluations that have accompanied these terms in the coercive power literature. Punishing is not always "bad" just as rewarding is not always "good." Coercive power can be used for beneficial purposes, such as rectifying inequalities produced by unequal reward power. Unfortunately, however, the reward dependence of actors who are also engaged in social exchange makes it unlikely that coercion will be used to oppose exploitation. Both power and justice are on the side of the exploiters.

ACKNOWLEDGMENTS

I gratefully acknowledge the generous support of the National Science Foundation throughout this entire research program (grants SES-8419872, SES-8921431, and SES-9210399). I also thank the many students who assisted in conducting the experiments, especially Suni Lee, Phillip Wiseley, and Theron Quist, each of whom worked for several years on the project. The editors of this volume provided helpful comments on the chapter. A book-length manuscript on this research program is forthcoming from Cambridge University Press.

NOTES

1. These early studies mistakenly concluded that punishment produces only temporary effects on behavior by arousing an emotional state in the organism. Contemporary research refutes these conclu-

sions, suggesting instead that punishment is highly effective, that it influences behavior through a process directly parallel to reinforcement, and that it produces positive side effects as often as negative ones (for reviews, see Axelrod and Apsche 1983; Azrin and Holz 1966; Van Houten 1983).

2.   Molm and Cook (1995) combine conditions 4 and 5 in a single scope condition; I separate them here for clarity.

3.   Unlike an exchange of material goods, performing a behavior that produces valued outcomes for another involves no actual transfer of resources and does not deplete the actor's supply of resources. Such behaviors may entail investment costs or intrinsic costs, of course; but to simplify the analysis, I assume they involve only opportunity costs.

4.   Analyses of data from this research program support the magnitude of loss aversion reported by decision theorists. Using a combined data set created from the first five experiments in the program, I examined the effects of experienced rewards (gains) and punishments (losses) on subjects' affective evaluations of their exchange partners' behavior (Molm 1991). Averaged across different power positions, the negative regression coefficients for punishment were 2.3 times the positive regression coefficients for reward.

5.   Some theorists restrict coercion to *threatened* sanctions, arguing that if the threat must be carried out, coercion has failed (e.g., Bachrach and Baratz 1963; Willer and Markovsky 1993). Other theorists specify other restrictions, for example, the threat or use of severe deprivation, physical violence, or loss of life (e.g., Bachrach and Baratz 1963; Bierstedt 1950; Dahl 1957). Because my aim is to compare structures of reward and punishment power that differ only on the sign of the sanction administered, I exclude such additional stipulations and omit threats (and promises) from my analysis.

6.   Emerson (1972b) originally defined power use by changes over time in the relative opportunity costs that A and B must incur. Thus, when A exercises a power advantage over B, A induces B to accept greater opportunity costs in return for A's exchange. This conception of power use is particularly appropriate for the reciprocal exchange relations I study. In reciprocal exchange, it is the potential to receive rewards *without* reciprocating that makes an asymmetrical exchange of rewards beneficial to the actor who rewards less often. Because that actor invests less time in the relation for a given level of returns, he or she has more opportunities to obtain rewards from other relations (i.e., the actor's opportunity costs are lower).

7.   Recently, Cook and Yamagishi (1972) have proposed a new algorithm for power-dependence theory that makes explicit point predictions for the division of exchange profit in negotiated exchanges in negatively-connected networks. Other theories that also make point predictions under similar conditions include elementary theory (Markovsky, Willer, and Patton 1988; Skvoretz and Willer 1993; Willer and Markovsky 1993) and expected value theory (Friedkin 1986, 1992).

8.   Reward withholding in reciprocal exchange is comparable to what Markovsky, Willer, and Patton (1988) call exclusion in negotiated exchange, as long as the withheld rewards are not obtained elsewhere.

9.   In this sense, strategic power use is comparable to Emerson's power-balancing mechanisms. Those mechanisms, which include network extension and coalition formation, are also not structurally induced but represent behavioral solutions to unsatisfactory exchange outcomes. The two are different, however, in the sense that strategic power use operates within an existing structure, while power-balancing mechanisms change the structure itself. Lawler (1992) refers to these as power-use tactics and power-change tactics, respectively, and Leik (1992) uses the terms strategic action and strategic agency.

10.   Both structurally and behaviorally, the *POs* were designed to be available, reliable, and reciprocating partners. Their structural relations with A and B—the real subjects—were always balanced or slightly imbalanced in favor of A or B, and their programmed behavior followed a modified tit-for-tat strategy: they never initiated punishment, occasionally initiated rewarding, and mainly reciprocated the subject's prior behavior, but at probabilities of 0.8 or 0.9, rather than 1.0, to prevent suspicion. All probabilities were randomized over blocks of 10 actions of the specified type by the subject.

11.   While coercive strategies are typically purposive, they are sometimes shaped without aware-ness, that is, when punishing actions are elicited emotionally and then maintained by their conse-quences. This process seems to explain the rapid development of coercive strategies employed by young children (see Patterson 1982) who quickly learn to cry, whine, or throw tantrums to increase adults' attention.

12.   This theorem does not imply that punishment power advantage structurally induces the use of coercive power; as I have argued, it does not. Punishment power advantage does reduce the potential loss from the partner's punitive retaliation, however, and thus reduces the risk of using coercive power strategically.

13.   Jasso (1980), Markovsky (1985), and others have developed mathematical formulations that model the justice process more precisely. My aim here is not to develop a formal theory of justice but to make explicit the assumptions I use to derive predictions about the perceived injustice of a partner's power use and reactions to those perceptions.

14.   Alternatively, actors might constrain their power use because of the moral weight of justice norms (Blau 1964) or because of a preference for acting fairly (Kahneman, Knetsch, and Thaler 1986). A substantial body of research suggests, however, that those who benefit from inequalities in outcomes or procedures feel little emotional distress (e.g., Hegtvedt 1990; Leventhal and Anderson 1970; Michaels, Edwards, and Acock 1984). Thus, actors are more likely to reduce their power use because it violates their *partners'* standards of fairness rather than their own.

15.   Perceptions of injustice were measured with a three-item scale (fair/unfair, just/unjust, equi-table/inequitable) with an alpha reliability of 0.81.

REFERENCES

Anderson, B., and D. Willer. 1981. "Introduction." Pp. 1-21 in *Networks, Exchange and Coercion: The Elementary Theory and Its Applications*, edited by D. Willer and B. Anderson. New York: Elsevier.

Axelrod, S., and J. Apsche. Eds. 1983. *The Effects of Punishment on Human Behavior.* New York: Aca-demic Press.

Azrin, N. H., and W. C. Holz. 1966. "Punishment." Pp. 380-447 in *Operant Behavior: Areas of Research and Application*, edited by W. K. Honig. New York: Appleton.

Bacharach, S. B., and E. J. Lawler. 1981. *Bargaining: Power, Tactics, and Outcomes.* San Francisco: Jossey-Bass.

Bachrach, P., and M. S. Baratz. 1963. "Decisions and Nondecisions: An Analytical Framework." *Amer-ican Political Science Review* 57: 641-51.

Berger, J., B. P. Cohen, and M. Zelditch, Jr. 1972. "Structural Aspects of Distributive Justice: A Status Value Formation." Pp. 119-146 in *Sociological Theories in Progress,* Vol. 2, edited by J. Berger, M. Zelditch, Jr., and B. Anderson. Boston: Houghton Mifflin.

Bierstedt, R. 1950. "An Analysis of Social Power." *American Sociological Review* 15: 730-38.

Blalock, H. M. 1987. "A Power Analysis of Conflict Processes." Pp. 1-40 in *Advances in Group Pro-cesses*, Vol. 4, edited by E. J. Lawler and B. Markovsky. Greenwich, CT: JAI Press.

_____. 1989. *Power and Conflict: Toward a General Theory.* Newbury Park, CA: Sage.

Blau, P. M. 1964. *Exchange and Power in Social Life.* New York: Wiley.

Boulding, K. E. 1969. "Toward a Pure Theory of Threat Systems." Pp. 285-292 in *Political Power: A Reader in Theory and Research*, edited by R. Bell, D. V. Edwards, and R. H. Wagner. New York: Free Press.

Burgess, R. L., and J. M. Nielsen. 1974. "An Experimental Analysis of Some Structural Determinants of Equitable and Inequitable Exchange Relations." *American Sociological Review* 39: 427-443.

Cook, K. S. 1975. "Expectations, Evaluations, and Equity." *American Sociological Review* 40: 372-88.

Cook, K. S., and R. M. Emerson. 1978. "Power, Equity and Commitment in Exchange Networks." *American Sociological Review* 43 :721-39.

Cook, K. S., R. M. Emerson, M. R. Gillmore, and T. Yamagishi. 1983. "The Distribution of Power in Exchange Networks: Theory and Experimental Results." *American Journal of Sociology* 89: 275-305.

Cook, K. S., and K. A. Hegtvedt. 1986. "Justice and Power: An Exchange Analysis." Pp. 19-41 in *Justice in Social Relations*, edited by H. W. Bierhoff, R. L. Cohen, and J. Greenberg. New York: Plenum Press.

Cook, K. S., and T. Yamagishi. 1992. "Power in Exchange Networks: A Power-Dependence Formulation." *Social Networks* 14: 245-265.

Dahl, R. A. 1957. "The Concept of Power." *Behavioral Science* 2 :201-5.

Deutsch, M. 1973. *The Resolution of Conflict*. New Haven, CT: Yale University Press.

Emerson, R. M. 1962. "Power-Dependence Relations." *American Sociological Review* 27: 31-41.

_____. 1972a. "Exchange Theory, Part I: A Psychological Basis for Social Exchange." Pp. 38-57 in *Sociological Theories in Progress*, Vol. 2, edited by J. Berger, M. Zelditch, Jr., and B. Anderson. Boston: Houghton-Mifflin.

_____. 1972b. "Exchange Theory, Part II: Exchange Relations and Networks." Pp. 58-87 in *Sociological Theories in Progress*, Vol. 2, edited by J. Berger, M. Zelditch, Jr., and B. Anderson. Boston: Houghton-Mifflin.

_____. 1981. "Social Exchange Theory." Pp. 30-65 in *Social Psychology: Sociological Perspectives*, edited by M. Rosenberg and R. H. Turner. New York: Basic Books.

Estes, W. K. 1944. "An Experimental Study of Punishment." *Psychological Monographs* 57: 3, Whole No. 263.

Etzioni, A. 1968. *The Active Society: A Theory of Societal and Political Processes*. New York: Free Press.

Fishburne, P. C., and G. A. Kochenberger. 1979. "Two-Piece Von Neumann-Morgenstern Utility Functions." *Decision Sciences* 10: 503-18.

Friedkin, N. 1986. "A Formal Theory of Social Power." *Journal of Mathematical Sociology* 12: 103-126.

_____. 1992. "An Expected Value Model of Social Power: Predictions for Selected Exchange Networks." *Social Networks* 14: 213-229.

_____. 1993. "An Expected Value Model of Social Exchange Outcomes." Pp. 143-167 in *Advances in Group Processes*, Vol. 2, edited by E. J. Lawler. Greenwich, CT: JAI Press.

Gouldner, A. W. 1960. "The Norm of Reciprocity: A Preliminary Statement." *American Sociological Review* 25: 161-78.

Gray, L. N. and I. Tallman. 1987. "Theories of Choice: Contingent Reward and Punishment Applications." *Social Psychology Quarterly* 50: 16-23.

Heath, A. F. 1976. *Rational Choice and Social Exchange: A Critique of Exchange Theory*. London: Cambridge University Press.

Hegtvedt, K. A. 1990. "The Effects of Relationship Structure on Emotional Responses to Inequity." *Social Psychology Quarterly* 53: 214-28.

Hegtvedt, K. A., and B. Markovsky. 1995. "Justice and Injustice." Pp. 257-280 in *Sociological Perspectives on Social Psychology*, edited by K. S. Cook, G. A. Fine, and J. S. House. Boston: Allyn and Bacon.

Hershey, J. C., and P. J. H. Schoemaker. 1980. "Risk Taking and Problem Context in the Domain of Losses: An Expected Utility Analysis." *Journal of Risk and Insurance* 47: 111-32.

Homans, G. C. [1961].1974. *Social Behavior: Its Elementary Forms*. New York: Harcourt Brace and World.

Jasso, G. 1980. "A New Theory of Distributive Justice." *American Sociological Review* 45: 3-32.

Kahneman, D., J. L. Knetsch, and R. H. Thaler. 1986. "Fairness and the Assumptions of Economics." *Journal of Business* 59: S285-S300.

_____. 1991. "The Endowment Effect, Loss Aversion, and Status Quo Bias." *Journal of Economic Perspectives* 5: 193-206.

Kahneman, D., and A. Tversky. 1979. "Prospect Theory: An Analysis of Decision Under Risk." *Econometrica* 47: 263-291.

———. 1982. "The Psychology of Preferences." *Scientific American* 246: 160-173.

———. 1984. "Choices, Values, and Frames." *American Psychologist* 39: 341-350.

Lawler, E. J. 1986. "Bilateral Deterrence and Conflict Spiral: A Theoretical Analysis." Pp. 107-30 in *Advances in Group Processes,* Vol. 3, edited by E. J. Lawler. Greenwich, CT: JAI Press.

———. 1992. "Power Processes in Bargaining." *The Sociological Quarterly* 33: 17-34.

Lawler, E. J., and S. B. Bacharach. 1987. "Comparison of Dependence and Punitive Forms of Power." *Social Forces* 66: 446-462.

Lawler, E. J., and J. Yoon. 1993. "Power and the Emergence of Commitment Behavior in Negotiated Exchange." *American Sociological Review* 58: 465-481.

———. 1996. "Commitment in Exchange Relations: Test of a Theory of Relational Cohesion." *American Sociological Review* 61: 89-108.

Leik, R. K. 1992. "New Directions for Network Exchange Theory: Strategic Manipulation of Network Linkages." *Social Networks* 14: 309-323.

Leventhal, G. S., and D. Anderson. 1970. "Self-Interest and the Maintenance of Equity." *Journal of Personality and Social Psychology* 15: 57-62.

Markovsky, B. 1985. "Toward a Multilevel Distributive Justice Theory." *American Sociological Review* 50: 822-39.

Markovsky, B., D. Willer, and T. Patton. 1988. "Power Relations in Exchange Networks." *American Sociological Review* 53: 220-36.

Michaels, J. W., J. N. Edwards, and A. C. Acock. 1984. "Satisfaction in Intimate Relationships as a Function of Inequality, Inequity, and Outcomes." *Social Psychology Quarterly* 47: 347-57.

Michaels, J. W., and J. A. Wiggins. 1976. "Effects of Mutual Dependency and Dependency Asymmetry on Social Exchange." *Sociometry* 39: 368-76.

Molm, L. D. 1981. "The Conversion of Power Imbalance to Power Use." *Social Psychology Quarterly* 16: 153-166.

———. 1985. "Relative Effects of Individual Dependencies: Further Tests of the Relation Between Power Imbalance and Power Use." *Social Forces* 63: 810-837.

———. 1988. "The Structure and Use of Power: A Comparison of Reward and Punishment Power." *Social Psychology Quarterly* 51: 108-122.

———. 1989a. "An Experimental Analysis of Imbalance in Punishment Power." *Social Forces* 68: 178-203.

———. 1989b. "Punishment Power: A Balancing Process in Power-Dependence Relations." *American Journal of Sociology* 94: 1392-1418.

———. 1990. "Structure, Action, and Outcomes: The Dynamics of Power in Exchange Relations." *American Sociological Review* 55: 427-447.

———. 1991. "Affect and Social Exchange: Satisfaction in Power-Dependence Relations." *American Sociological Review* 56: 475-493.

———. 1993. When Coercive Power Fails: Incentive and Risk in Social Exchange." Presented at the annual meetings of the American Sociological Assocation, Miami, FL.

———. 1994a. "Is Punishment Effective? Coercive Strategies in Social Exchange." *Social Psychology Quarterly* 57: 75-94.

———. 1994b. "Dependence and Risk: Transforming the Structure of Social Exchange." *Social Psychology Quarterly* 57: 163-176.

Molm, L. D., and K. S. Cook. 1995. "Social Exchange and Exchange Networks." Pp. 209-235 in *Sociological Perspectives on Social Psychology,* edited by K. S. Cook, G. A. Fine, and J. S. House. Boston: Allyn and Bacon.

Molm, L. D., T. M. Quist, and P. A. Wiseley. 1993. "Reciprocal Justice and Strategies of Exchange." *Social Forces* 72: 19-43.

———. 1994. "Imbalanced Structures, Unfair Strategies: Power and Justice in Social Exchange." *American Sociological Review* 59: 98-121.

Morgan, P. M. 1977. *Deterrence: A Conceptual Analysis.* Beverly Hills, CA: Sage.

Patterson, G. R. 1982. *Coercive Family Process.* Eugene, OR: Castalia Publishing.

Pruitt, Dean G. 1981. *Negotiation Behavior.* New York: Academic Press.

Rubin, J. Z., and B. R. Brown. 1975. *The Social Psychology of Bargaining and Negotiation.* New York: Academic Press.

Schelling, T. C. 1960. *The Strategy of Conflict.* New York: Oxford University Press.

Skinner, B. F. 1938. *The Behavior of Organisms.* New York: Appleton.

Skvoretz, J., and D. Willer. 1993. "Exclusion and Power in Exchange Networks." *American Sociological Review* 58: 801-818.

Stolte, J. F. 1983. "The Legitimation of Structural Inequality: The Reformulation and Test of the Self-Evaluation Argument." *American Sociological Review* 48: 331-42.

_____. 1987. "Legitimacy, Justice, and Productive Exchange." Pp. 190-208 in *Social Exchange Theory,* edited by K. S. Cook. Newbury Park, CA: Sage.

Tedeschi, J. T., B. R. Schlenker, and T. V. Bonoma. 1973. *Conflict, Power, and Games.* Hawthorne, NY: Aldine.

Thibaut, J. W., and H. H. Kelley. 1959. *The Social Psychology of Groups.* New York: Wiley.

Thorndike, E. L. 1932. *The Fundamentals of Learning.* New York: Teachers College, Columbia University.

Tversky, A., and D. Kahneman. 1991. "Loss Aversion in Riskless Choice: A Reference-Dependent Model." *The Quarterly Journal of Economics* 106: 1039-1061.

Van Houten, R. 1983. "Punishment: From the Animal Laboratory to the Applied Setting." Pp. 13-44 in *The Effects of Punishment on Human Behavior,* edited by S. Axelrod and J. Apsche. New York: Academic Press.

Willer, D. 1981. "The Basic Concepts of Elementary Theory." Pp. 25-53 in *Networks, Exchange, and Coercion: The Elementary Theory and Its Applications,* edited by D. Willer and B. Anderson. New York: Elsevier.

_____. 1987. *Theory and the Experimental Investigation of Social Structures.* New York: Gordon and Breach Science Publishers.

Willer, D., and B. Anderson. Eds. 1981. *Networks, Exchange, and Coercion: The Elementary Theory and Its Applications.* New York: Elsevier.

Willer, D., and B. Markovsky. 1993. "Elementary Theory: Its Development and Research Program." Pp. 323-363 in *Theoretical Research Programs: Studies in the Growth of Theory,* edited by J. Berger and M. Zelditch, Jr. Stanford, CA: Stanford University Press.

Wrong, D. H. 1979. *Power: Its Forms, Bases and Uses.* New York: Harper and Row.

Yamaguchi, K. 1996. "Power in Networks of Substitutable/Complementary Exchange Relations: A Rational-Choice Model and an Analysis of Power Centralization." *American Sociological Review* 61: 308-332.

UNDERSTANDING THE POSITIVE CONSEQUENCES OF PSYCHOSOCIAL STRESSORS

Michael J. Shanahan and Jeylan T. Mortimer

ABSTRACT

Stress can promote both disease and growth. Drawing on Selye's biological model and relevant empirical research, we propose a conceptual model of psychosocial eustress which explains how social stressors result in positive outcomes, including enhanced self-conceptions, affect, and physiological change. Primary eustress refers to processes whereby eustressful outcomes are experienced simultaneously with stressors. The individual is goal-oriented, pursues tasks with high levels of involvement, and receives feedback indicating progress in neutralizing the stressor. Secondary eustress refers to processes whereby eustressful outcomes occur because earlier stressful experiences have heightened the individual's adaptive capacity in later stressful encounters. Mechanisms of secondary eustress include the raising of thresholds of reactivity, strengthening of motivational structures, and promoting selection into contexts which are more challenging but still controllable.

Advances in Group Processes, Volume 13, pages 189-209.
Copyright © 1996 by JAI Press Inc.
All rights of reproduction in any form reserved.
ISBN: 0-7623-0005-1

This essay sketches a social psychological model by which the stress process engenders adaptation. The interrelations of stress and the self have been given central attention in the classic and contemporary literatures of sociological and psychological social psychology (Zimmerman 1988). While contemporary social psychologists specializing in stress research have indicated salutary and detrimental consequences of stressors for the self (e.g., Thoits 1994; Turner and Avison 1992), the processes underlying positive consequences deserve further scrutiny.

Reviews of stress research frequently note the low magnitude of associations between putative stressors and distressful outcomes (cf. Turner, Wheaton, and Lloyd 1995). Yet, people are adaptive and, when threatened with stressors, make a wide range of adjustments to avoid distressful reactions. In the process of coping with stressors, the individual's adaptive capacity may be enhanced and, by maintaining the self's integrity in the wake of stressors, the individual may experience a range of positive outcomes. What are the social psychological mechanisms underlying these processes?

We begin by examining the biological and physical models which inspire psychosocial stress research. Both Hans Selye's biological model and Blair Wheaton's engineering model recognize that the stress process often results in enhancement of the organism's or material's adaptive capacity. However, this possibility is relatively unexamined in psychosocial research. We then propose a social psychological model of eustress which emphasizes the motivational role of self-conceptions and the consequences of successful coping strategies. The positive consequences of stressors may arise simultaneously with the experience of stressors (primary eustress) or in subsequent stressful circumstances (secondary eustress).

Virtually all coping research examines how adaptive responses to stressors act as buffering or mediating mechanisms, diminishing the effects of potentially distressful experiences. Our focal question is different: how does successful coping in response to stressors increase adaptive capacity and promote positive psychosocial change? Our model-building strategy is to develop a comprehensive analogy between Selye's biological model and a psychosocial model, while also taking into account major empirical findings in the sociological study of stress.

STRESS AS AN ADAPTIVE PROCESS: BIOLOGICAL AND PHYSICAL MODELS

The Biological Stress Process: Selye's *The Stress of Life*

Stress is the basis for both human disease and development. This intriguing premise was recognized by the physician Hans Selye (1976) in his seminal volume, *The Stress of Life*. According to his biological model, the stress process is a particular syndrome (the General Adaptation Syndrome) which involves the adrenal, thymicolymphatic, and gastrointestinal systems as they pass through the three stages of alarm, resistance, and exhaustion. The stress process is nonspecifically-induced, meaning that all or most of the biological system is affected and that the process can be triggered by many different kinds of agents. Selye's fundamental contribution was to recognize that nonspecific damage can be identified in addition to the specific symptoms of any disease and that this nonspecific damage (distress) is due to the three-stage stress process.

From this biological perspective, any agent which initiates the stress process is a stressor. Stressors are ubiquitous and account for both human disease and growth. As Selye (1976, pp. 63-64) observes:

> The stress of exhilarating, creative, successful work is beneficial, while that of failure, humiliation, infection is detrimental. The stress reaction, just as energy consumption, may have good or bad effects. The statement of 'he is under stress' is just as meaningless as the expression 'he is running a temperature.'

Distress results when external agents deplete the adaptive resources of the biological system, although this is relatively rare. More common are biological reactions which neutralize stressors. In many instances, external agents activate alarm and resistance, but homeostasis (i.e., the maintenance of a neutral state) results. Either syntoxic corticoids mobilize the defense system to ignore or retreat from external agents or catatoxic corticoids isolate and destroy external agents. Internal (e.g., heredity) and external (e.g., diet) factors condition the effectiveness of these corticoids. Selye states that developmental adaptation occurs when pre-existing capacity is enhanced, while redevelopmental adaptation occurs when an organism must reconfigure to meet external challenges.

Much less understood is the eustressful reaction, which Selye (p. 466) defines as "a pleasant or curative stress." The General Adaptation Syndrome is capable of producing both distress and eustress (Selye 1975). In discussing the ethical implications of his medical model, Selye states that eustress is fulfillment due to an accumulation of "capital gain to meet future needs" (1976, p. 451). In this sense, eustress is similar to his biological concept of heterostasis. Occasionally, when faced with excessive demands, individuals are treated to raise the level of assault they can withstand. Heterostasis, the stimulation of adaptive capacity, can be found, for example, in immunizations against infectious disease. Selye offers no systematic treatment of eustress, but a plausible interpretation is that *biological eustress is developmental or redevelopmental adaptation resulting from an alarm-resistance reaction. Developmental adaptation refers to an increment in existing physiological structure or process which provides added capacity to withstand stressors. Redevelopmental adaptation refers to a reorganization of physiological structure or process which provides added capacity to withstand stressors.*

The Physical Stress Process: Wheaton's Engineering Model

In contrast to Selye's General Adaptation Syndrome, Wheaton (1994) argues for the utility of a stress model inspired by the engineering and material sciences. The engineering model defines stress as load exerted by force divided by the capacity of the material to resist. When the load exceeds resistance capacity, the force becomes a stressor leading to a distressful outcome—damage to the material. Context influences the response to the stressor: for example, heat alters the resistance capacity of materials. According to Wheaton, the engineering model allows for the enhancement of the material's resistance through changes in its elastic limit. Materials under increasing levels of stress will elongate or compress, achieving greater resistance: the engineering model "allows for enhancement of coping capacity as a result of facing stressful situations" (p. 10). Ultimately, however, materials have a finite elastic range and stress levels in excess of this range result in damage.

From this basic model, Wheaton specifies a psychosocial model of stress. He defines psychosocial stressors as "threat, demands, or structural constraints that, by the very fact of their occurrence or existence, call into question the operating integrity of the organism" (pp. 4-5). Although stressors lead to distressful results when they exceed elastic limit, they also exist at subthreshold levels of threat. For Wheaton, it is critical to understand the circumstances under which stressors lead to distress. He addresses this issue by emphasizing the importance of context and resistance capacity, themes highlighted in the engineering model.

First, context defines the meaning of stressors and must be taken into account in evaluating the degree of threat posed by a stressor. For Dohrenwend, Link, Kern, Shrout, and Markowitz (1990), context includes recent events, ongoing social situations, personal dispositions, and early life experiences. Elder, George, and Shanahan (1996) demonstrate the utility of life course theory for understanding the interrelations of context and person. Wheaton (1994, p. 48) advocates a biographical approach in which context is any situational factor that defines the meaning and potential of a stressor: "... we are surely specifying the meaning of a stressor when we specify factors that alter its interpretation, the course it takes, or the complexity of the situation it poses." Wheaton promotes a contextual approach to the assessment of meaning instead of asking about it directly. Cognitive appraisals of stressors raise the possibility of measurement confounds between the self-reported meaning of events and self-reported indicators of distress.

Second, in the engineering model, as force increases, resistance to the load increases up to the elastic limit of the material. However, Wheaton demonstrates that while the stress produced by the load diminishes over time, as resistance increases, its rate of decrease will be nonlinear and decelerating. *In terms of the engineering model, then, eustress is the enhancement of resistance early in expo-*

sure to forces, before the material's ultimate elastic limit to resist the load is reached.

Summary

Both the biological and engineering models were developed to explain distressful outcomes. Selye's model was formulated to explain the effects of noxious agents; the theory primarily details the process by which organisms reach biological exhaustion and incur physiological damage. The engineering model was formulated to understand how much pressure materials can withstand before structural damage results. Yet, both models allow for the strengthening of the organism or material from stressors. According to the biological model, stressors can heighten existing capacity to resist or can lead to the emergence of new adaptive capacities through reorganization, at least before exhaustion sets in. According to the engineering model, resistance is enhanced with early increments in stressors, before the ultimate elastic limit is reached (see also Smith 1987).

The possibility that stressors can promote adaptation and growth has not been prominent in sociology.[1] We propose a psychosocial model of eustress which preserves Selye's fundamental contribution: the stress process is a particular syndrome which is nonspecifically induced and which can result in nonspecific damage or growth. Eustressful results—adaptive consequences originating with the stress process— are nonspecifically induced because a wide range of phenomena can cause them. The process of eustress is characterized by an interruption (or incompletion) of Selye's stress process. The person experiences alarm (causing activation of self-relevant motives) and resistance (adaptive strategies). However, exhaustion does not occur. Instead, there is a nonspecific improvement in functional capacity, possibly involving positive increments in self-concept, affect, or physical health. We turn now to a consideration of these issues.

A PSYCHOSOCIAL MODEL OF EUSTRESS:
THE SELF-SYSTEM AND ADAPTIVE STRATEGIES

Selye argued that biological distress is a relatively rare reaction to stressors. Perhaps a similar observation is true for psychosocial distress. Turner and Avison (1992) find that over half of all negative life events are resolved successfully in a community sample of physically limited individuals. The rate of success does not vary significantly across types of events (e.g., financial and work-related, death of someone close, interpersonal conflicts). Using a different measurement strategy, Thoits (1994) reports that over half of all work problems are solved (although unsolved problems in love and marriage are much more common). A model of eustress represents the successful social psychological response to stressors. The basic model is shown in Figure 1, which we discuss in the remainder of the paper.

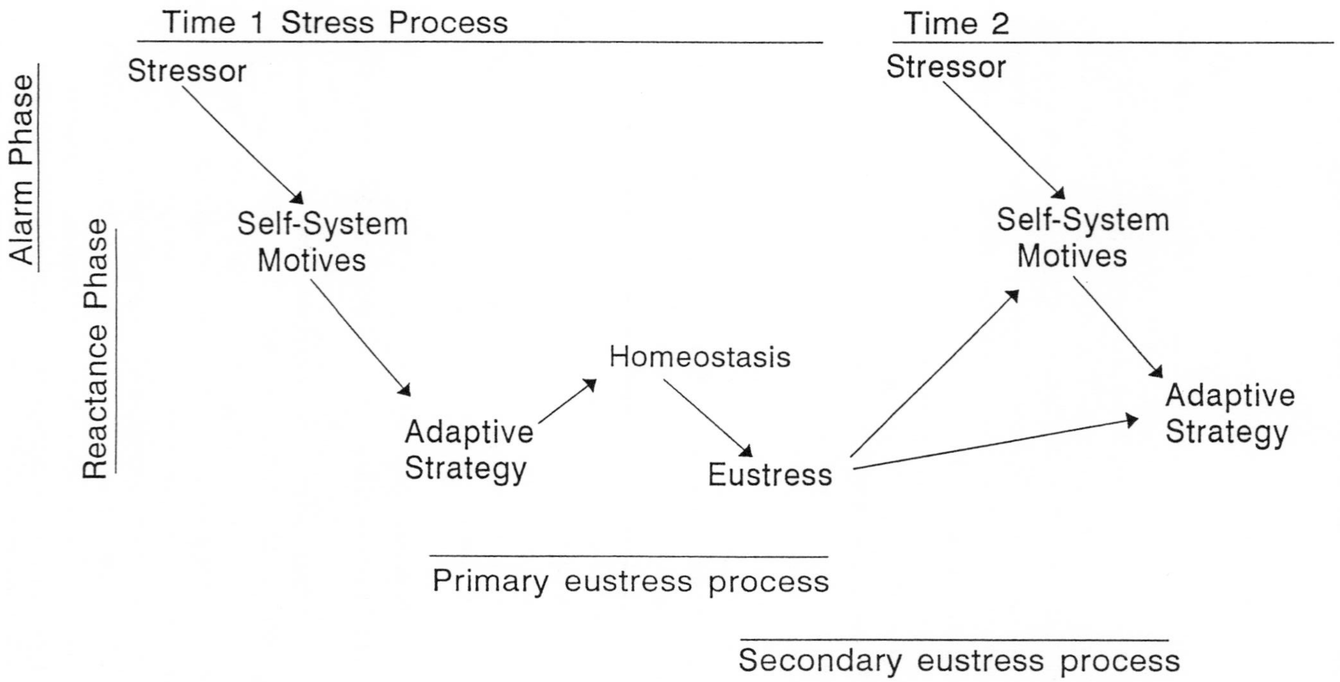

Figure 1. Basic Model of Psychosocial Stress With Eustressful Outcome

Defining Psychosocial Stressors: The Alarm Phase

An impressive body of empirical research links structurally-based stressors to distress by way of self-efficacy and related concepts like fatalism (Aneshensel 1992; Mirowsky and Ross 1989). For instance, Wheaton (1980) proposes a model whereby control attributions moderate the effects of problems or stressors on adaptation potential. He observes that people of lower socioeconomic status (considered less successful by normative standards) have lower levels of internal control and engage in fewer coping efforts. In subsequent research, drawing on these ideas and the work of Kohn, Wheaton (1983) demonstrates that fatalism and intellectual flexibility moderate the impact of acute stressors on depressive and schizophrenic symptoms. In short, stressors are more apt to cause distress when low efficacy restricts coping efforts.

Critical to this approach is the motivational role of efficacy, a subject of much research in psychology. Bandura (1993) explains how self-efficacy operates as a motivating factor in terms of attribution theory (for an application of this approach to stress research, see Wheaton 1980), expectancy-value theory, and goal theory. His program of experimental research indicates that self-efficacy predicts adaptive strategies, which in turn influence performance levels (see also DeCharms and Muir 1978).

Other models similarly emphasize the motivational role of the self-system in adaptation (Markus and Wurf 1987). As Gecas (1986) argues, "by virtue of having a self-concept the individual is motivated to maintain and enhance it, to conceive of it as efficacious and consequential" (p. 138). Self-concepts with motivational potential also include self-esteem and authenticity. Gecas notes that the latter self-conception has been difficult to define, but he refers to it as "the meaning and significance of what one does" (p. 141); similarly, Erikson (1993) posits that individuals feel authentic when they address the self's core commitments (see Kiecolt 1994; Turner and Billings 1991).

Drawing on these ideas, the model of psychosocial stress posits that socially-based challenges to one's self-system are the psychosocial analog to Selye's alarm function: psychosocial stressors are any phenomena which instill alarm by activating one's motives to maintain or enhance efficacy, esteem, and authenticity.[2] The social psychological mechanisms of the alarm phase have been elucidated by Kiecolt (1994), who examines connections between stressors and decrements in self-conceptions. Social psychological mechanisms which link stressors with lowered efficacy, esteem, or authenticity (i.e. the alarm phase) include unfavorable reflected appraisals, social comparisons, and self-perceptions.

A useful distinction can be made between the alarm threshold and the motivational structure which is activated. That is, individuals with strongly positive self-conceptions are less likely to be alarmed or to experience phenomena as stressors. Efficacious individuals are less apt to experience a minor challenge as a threat to their self-efficacy. Once alarmed, however, individuals may be more or less motivated to mobilize themselves in response to stressors.

This psychosocial approach is consistent with Selye's strategy of defining a stressor as anything which initiates the stress process. Thus, what may constitute a stressor for some persons, may not be stressors for others. For example, Wheaton (1990) shows that divorce can actually be a stress relief. In this case, characteristics of the marriage itself trigger the self's motives (i.e., the marriage is a source of stressors), while the divorce does not (i.e., the divorce is not a stressor). According to this conceptualization, phenomena can be classified as stressors only if they in fact initiate the alarm phase by challenging the integrity of the person's self-conceptions. Ultimately, this is an issue of meaning.

The Reactance Phase and Eustressful Outcomes

If psychosocial stressors trigger alarm by activating the self-system's motivations, how can the psychosocial reactance phase be characterized? The psychosocial reactance phase is comprised of adaptive strategies, including coping efforts and the elicitation of social support. At this point in the discussion, specifically what the individual does is not important; a wide range of coping strategies have been identified. Particular strategies are differentially available depending on one's placement in the social system, and they vary in effectiveness (e.g., Menaghan and Merves 1984).

Once the self-system's motives initiate adaptive efforts, a number of consequences may ensue. Virtually all stress research focuses on distress, which follows from unsuccessful adaptive efforts. Homeostasis is the maintenance of one's self-system and reflects successful coping. Eustressful consequences occur when individuals succeed in meeting the challenges of stressors. The maintenance of the self-system (homeostasis) through successful coping with stressors leads to eustressful outcomes, which may include an enhancement of the self-system, positive affect and improvements in physical health.

For example, an increment in efficacy can occur by way of self-perception. Attribution theorists focus on the person's understanding of the causes of events and the implications of these interpretations for subsequent cognitions and behaviors. Bem (1972, p. 40) emphasizes that the actor's cognitions give meaning to the situation as well as affect the behavioral response to it. Just as through social perception persons infer motives and intentions from the behavior of others (Heider 1958, chapter 2), they also make inferences about their own dispositions by observing their behaviors in given contexts (Bem 1972). This is especially likely when "internal cues" are ambiguous, that is, when attitudes, intentions, or beliefs are weakly defined (Ross and Fletcher 1985, pp. 98-99). Thus, upon observing their own behavior which results in successful coping, actors attribute efficacy to themselves.

But eustressful results can include a range of positive outcomes, such as positive affect and possibly even physical health. Lazarus (1991) proposes a theory of emotions which classifies affective states into two broad categories: those involving goal incongruence (e.g., anger or guilt-shame) and goal congruence (e.g., happiness/joy, pride, and relief). The latter feelings arise when on-going activity is appraised as congruent with the actor's goals. By implication, when individuals can efficaciously pursue their goals in the wake of challenges, a variety of positive emotions result. Over time, these various positive emotional experiences enhance subjective well-being, a more general appraisal of the success of one's personal life agenda.

Whereas Lazarus notes that very little solid evidence links positive emotional states with improvements in somatic health, he raises the possibility that positive (goal congruent) emotions can act as "restorers," permitting healing and recovery. Edwards and Cooper (1988) also suggest that positive psychological states can improve physical health. Karasek, Russell, and Theorell (1982) argue that situations involving both high challenge and high control are associated with increments in physiological functioning. They speculate that these situations produce anabolic hormones which, when in excess of catabolic hormones, can lead to multifaceted improvements in health.

Both primary eustress and secondary eustress are identified in Figure 1. *Primary eustress* is a successfully managed alarm-resistance sequence as it occurs in on-going activity. *Secondary eustress* refers to processes which render otherwise distressful circumstances innocuous because prior eustressful experiences have enhanced adaptive capacity. Briefly, secondary eustress includes several different mechanisms: the enhancement of self-concepts (thereby raising thresholds of reactivity to potential stressors), the strengthening of motivational structures, and selection into more challenging, but controllable, contexts. The mechanisms involving thresholds of reactivity and selection result in an avoidance of potentially distressful stressors; this avoidance can produce eustressful results by way of self-appraisals. That is, when individuals view themselves as able to avoid distress, positive self-conceptions and affect are likely to result. Common to both primary and secondary eustress are increments in functional capacity due to the successful coping with (or response to) stressors. The two differ in their temporal frames.

The model shares common themes with Turner and Avison's (1992) empirical study of resolved and unresolved life events. Drawing on Erikson's crisis theory, they argue that only unresolved negative life events produce distress. Indeed, their research offers empirical support for this hypothesis: unresolved events, but not resolved events, increase depressive symptoms. In a limited number of instances (those involving deaths of family members or friends and financial/work-related events) successfully resolved events significantly and *negatively* predict depressive symptoms.

Similarly, Thoits (1994) examines the role of solved vs. unsolved events. She argues that stress research has failed to consider the individual as motivated to avoid distressful outcomes. Indeed, she observes that psychological distress is predicted by unsuccessful problem-solving attempts but not by solved work or love/marriage problems. Her findings also suggest that low mastery and esteem select individuals into stressful circumstances, but at the same time these self-conceptions are enhanced by successful problem-solving outcomes.

Summary

According to the proposed theoretical model, shown in Figure 1, psychosocial stressors are any phenomena which trigger alarm by activating the individual's self-system's motives; psychosocial stressors are nonspecific in that a wide range of socially-based experiences can present a challenge to the self's esteem, efficacy, or authenticity. The challenge occurs by way of unfavorable social comparisons, self-perceptions, or reflected appraisals (Kiecolt 1994). The self's motives prompt the reactance phase, which consists of adaptive efforts (the elicitation of social supports and a wide range of coping strategies) to maintain or enhance the self. Psychosocial distress results when ineffective responses to stressors lead to an exhaustion of psychosocial resources, with its concomitant negative outcomes. Homeostasis, the maintenance of the self-system in the face of challenges, results from effective coping and produces eustressful results—increments in functional capacity due to an alarm-resistance reaction. Nonspecific eustressful reactions include a range of outcomes, such as a heightening of efficacy, esteem, or authenticity; positive affect; and possibly improved health.

How are stressful situations structured so as to promote primary eustress in the immediacy of experience? What particular social psychological processes connect prior successful coping with later stressful experiences? We turn now to a consideration of these issues by examining primary and secondary processes of eustress.

MECHANISMS OF PSYCHOSOCIAL STRESS AND ADAPTATION: PRIMARY EUSTRESS

Primary eustress occurs when there is engagement in challenging problems in anticipation of eventual success. If there is perceived progress in overcoming the difficulty, or if the effort as a whole is successful, the person comes to see herself as capably coping with adverse circumstance. This attitude may be generalized over time and incorporated into the self, engendering heightened senses of efficacy, authenticity, and esteem. These processes are elucidated in Mead's classic conceptualization of the act as well as Csikszentmihalyi's more recent analysis of flow experiences.

Mead's Philosophy of the Act and the Aesthetic Experience

Mead's philosophy of the act is especially relevant to a consideration of eustress given its pragmatic premise: acts begin with problematic situations, a "lack of adjustment between the individual and the world" (Mead 1938, p. 6), and end with successful adaptation. For Mead, the social act is comprised of four stages: impulse, perception, manipulation, and consummation. Impulse represents the initial process of becoming aware of a problematic situation, based on instinct or prior experience. Perception refers to the thoughts surrounding the interpretation of this response. In the stage of manipulation, the actor mentally rehearses alternative lines of action to deal with the problem. This stage involves viewing the self as one object among many in the situation. For action to be successful, the self must put itself in the place of other objects and imagine their effective coordination.

The actor empathically takes the roles of other persons, but the productive act also involves taking the place of non-human objects so as to appreciate the physical laws governing the situation. Thus, the self becomes an object and then considers the interrelationships between all of the objects (self, other actors, raw materials, tools) necessary to successfully execute a response to the problematic situation. By "taking on" these objects, the self becomes aware of its role in dealing with the problem. The consummatory phase encompasses the act's completion. The self orchestrates objects, including itself, in the adaptive process.

The aesthetic experience consists of contemplating the consummatory phase during the process of manipulation. It includes cognitive states and emotive affects prompted during the action of manipulation and focused on completion. Mead (1926, p. 387) writes, "The enjoyment of its [the labor's product] ultimate use must be suggested by the intermediate steps in its production, and must flow naturally into the skill which constructs it."[3]

All social acts are situated in the present, directed toward the future, and draw heavily from the past. Manipulation suggests consummation given the "reliable uniformities and laws of affairs" that the past hands to the present. To the extent that past experiences are useful, manipulation is followed by consummation. But when past experience inadequately informs the situation, uncertainty prevents effective action. We may infer, then, that distress is produced when the act cannot continue to completion with some threshold level of perceived certainty. The situation retains its problematic character, and the actor views his or her self as inefficacious.

However, Mead also believed that problem-solving activity can be a creative process and that the aesthetic experience is heightened when new causal patterns are discovered and when activities are novel and expressive of new aesthetic constructions. In such circumstances, the act requires creative synthesis of the past's uniformities. Thus, the experiences of novelty and creative synthesis, which might be expected to be particularly "eustressful," as well as reliability and uniformity, are basic to the aesthetic experience.

Csikszentmihalyi's Flow and Optimal Experience

Csikszentmihalyi (1990) elucidates the psychological dynamics of eustress. Conceptually related to Mead's aesthetic experience are his concepts of "flow" and "optimal experience." Central to the optimal experience is a sense of challenge. The optimal experience involves clearly understood goals in response to such challenge, deep concentration in goal-directed activities, the ability to exercise control, and the receipt of feedback indicating progress.[4] Flow is a psychological state which defines the optimal experience: psychic energy invested in activity flows effortlessly with complete concentration on the task. Paradoxically, even though self-consciousness is minimized, the self emerges from the optimal experience as a more complex entity. Feelings, intentions, goals, and senses become more highly integrated.

Csikszentmihalyi observes that virtually any stressor can lead to an optimal experience. Even major catastrophes may bring with them the opportunity for growth, provided the individual has a sense that "skills are adequate to cope with the challenge at hand, in a goal-directed, rule-bound action system that provides clear clues as to how well one is performing" (p. 71). While many stressors represent challenges and people do initiate focused activity in response, feedback indicating progress toward the resolution of the problem may be absent. Such situations would eventuate in distress.

Summary

Primary eustress occurs in the immediacy of successful coping efforts and includes a wide range of positive outcomes. Common to both Mead's aesthetic experience and Csikszentmihalyi's optimal experience, a eustressful process involves engagement in an activity in the context of a challenge (stressor). The actor is goal-oriented, pursues the goal with maximal involvement, and receives feedback indicating progress toward the goal. If the individual's mental and physical efforts are sufficient, given the task's demand characteristics, eustress results. The authentic self is likely to be expressed, since situations characterized by challenge promote feelings of self-authenticity (Turner and Billings 1991). The success of one's efforts is perceived and self-conceptions are enhanced, particularly self-efficacy. In contrast, challenges which offer little feedback about progress toward a goal are likely to cause distress.

MECHANISMS OF PSYCHOSOCIAL STRESS AND ADAPTATION: SECONDARY EUSTRESS

A model of secondary eustress explains how previously eustressful experiences can enhance functional capacity in subsequent, potentially stressful circumstances. What are the social psychological mechanisms which describe this more temporally-extended process? We focus on social motivation and learning-generalization, although our discussion is only a sketch given the extensive literature on these topics. Three major social psychological mechanisms of secondary eustress can be identified: (1) raising domain-specific and global self-conceptions, thresholds of reactivity to potential stressors ("steeling"), (2) strengthening the motivational structure so that, following the experience of alarm, motives to actively deal with stressors are stronger, and (3) promoting selection into contexts which are more challenging but still controllable.

Efficacious Functioning and Motivational Structures

Theories of motivation are particularly relevant to a dynamic model of secondary eustress since they contribute to an understanding of how previous eustressful experiences affect later coping. An earlier stress process can affect subsequent stress processes in several ways; motivation is especially relevant since past experiences lead to enhancements or decrements in the self-system's motivational structure.

If increments in efficacy result from successful coping with stress, there is an important implication for reactions to subsequent stressors. In his classic essay on competence motivation, White (1959) argues that individuals have a motive to enhance their sense of "achieved capacity," or competence. A stimulus which is challenging and which presents novel situations triggers the person's desire to feel efficacious. White argues that successful exercises of such "effectance motivation" result in a greater likelihood of efficacy-driven behaviors in the future. Efficacious experiences heighten one's sense of competence, increasing the likelihood of future competence-driven behaviors.

White's formulation suggests that individuals who successfully resolve stressors at Time 1 will bring heightened capacities to the stress process at Time 2. This heightened capacity results from higher levels of self-efficacy and motivation. When self-efficacy is enhanced, a higher threshold of reactivity is established: potential stressors are less likely to activate alarm. Moreover, a sense of self-efficacy heightens the individual's motivation to take risks and actively cope with stressors. Thus, success in dealing with stressors has a double payoff: higher thresholds of reactivity and stronger motivational structures.

A different approach is advanced by Atkinson and his colleagues (Atkinson 1964; Atkinson and Raynor 1974) but leads to similar expectations. Briefly, Atkinson's achievement motivation theory holds that individuals will engage in achievement-related behaviors as a function of their success motive, the probability of success, and the incentive value associated with success.[5] Especially relevant for our purposes, the probability of success is a subjective evaluation reflecting the expectancy of goal attainment. Research drawing on Atkinson's model demonstrates that the subjective evaluation of success can be heightened by past rewarding experiences. For example, students given unsolvable math problems are not likely to work on subsequent problems even when they are readily solvable; in contrast, students given solvable math problems move directly to the second set of problems. This suggests that when individuals experience successes in coping with stressors, they develop stronger motivation to engage in coping efforts given subsequent stressors. Past experiences have heightened the subjective probability of success, which increases the motive to approach success. However, not all past experiences have these beneficial results; future research should examine what types of Time 1 stressors affect the the subjective evaluation of success for types of Time 2 stressors.[6]

Raynor and his colleagues (Entin and Raynor 1973; Raynor 1969), in an interesting elaboration of this research, argue that achievement motivation reflects connections between the present and future. The probability of success (or failure) bears on one's present achievement motivation, but the meaning of resolving the current problem in a more broadly defined life-space is also important. Individuals are more motivated to overcome problems when they detect some probability of success *and* when the current problem fits into a plan of future acts, leading to a desired ultimate goal, which also has some probability of success. A contingent path—tasks which are serially related—produces a higher level of performance. Thus, stressors which challenge the individual not only in the present but which also have implications for valued future states are more likely to activate the self-system's motives and promote adaptive efforts.

Learning-Generalization of Adaptive Self-Concepts and Other Psychological Traits

Kohn and Schooler's (1983) "learning-generalization" model, also suggests the possibility of secondary eustress. Learning-generalization theorists argue that the self-concept and other psychological traits are altered as the person adapts to demanding circumstances. Individuals come to value those traits in themselves that will enable them to successfully adapt and select situations which influence their future adaptive capacity. Learning processes may thus increase the effectiveness of subsequent adaptive strategies. Learning-generalization emphasizes the malleable nature of the self, although, given its historical roots in behaviorism, this approach is just beginning to consider the possibility of a complex self (Caplan and Schooler 1992).

According to Thorndike's (1913) law of effect, the organism's behavioral tendencies in a given situation will be strengthened when the effects or consequences following enacted behaviors are satisfying. Learning thus consists of the strengthening of bonds between situations and behaviors. The organism possesses needs or drives which contribute to arousal, attends to the situational cues for behavior in the environment, acts, and the reward (or punishment) following such behavior explains the maintenance or extinction of the response (Hall and Lindzey 1957;

Lott and Lott 1985, p. 109). Thus, prior experiences of the person following particular modes of response predict subsequent behavior in the same situation.[7]

Because situational cues are not identical across successive occasions, generalization necessarily comes into play. Stressful life-events which are structurally similar (e.g., death of a spouse, death of a parent) may be responded to similarly if the first event was dealt with successfully. As Hall and Lindzey (1957, p. 434) observe, "habits learned in one situation will tend to be transferred or generalized to other situations to the extent that the new situations are similar..."

The Kohn-Schooler research program has focused attention on self-direction in the work environment. As Miller (1988, p. 341) makes clear, this emphasis is premised on the belief that complexity and autonomy, as well as their psychological correlates (interest, challenge, and a sense of responsibility), are sources of gratification or deprivation in the work setting. Using the language of behaviorism, she describes self-direction as "a fundamental reward of work and a stimulus to which the organism is attentive" (Miller 1988, p. 342).

Learning theorists are mainly concerned with observable behaviors and less interested in cognitive processes. Importantly, however, as in much other contemporary social psychological work (Lott and Lott 1985), the "learning-generalization" paradigm of work conditions and psychological functioning pairs concepts from the learning theoretic framework with cognitive processes and outcomes (Schooler 1987, pp. 24-25). The conceptualization of what is learned goes beyond behavior to include cognitions of self and others, values, and other psychological attributes (Breer and Locke 1965).

Particularly relevant to our present concerns, Kohn and Schooler (1983) find that occupational self-direction has significant impacts on key dimensions of self and stress: self-confidence and anxiety. These cognitive evaluations, elicited and rewarded by the work environment, are generalized beyond this context. Self-perception of successful and autonomous adaptation to the challenges posed by stressors will foster favorable conclusions about self-abilities; these conclusions are subject to generalizations across domains of action. They also motivate selection of environments that are consistent; self-confident individuals will take on challenges which further enhance their adaptive capacities and self-images. For these individuals, aesthetic experiences are more common and such episodes are linked by way of effectance motivation and related processes. Those who lack such confidence tend to withdraw from problematic situations. For these individuals, distress ideally is reduced or eliminated.

The learning-generalization model of social structure and personality focuses on the demands and opportunities posed by the environment; persons are expected to react similarly to them (Mortimer and Simmons 1978). However, models of social cognition (Schneider 1991) lead to the expectation that other environmental stressors condition the effects of stressors at work. They posit that hierarchically-organized schemata, which guide attention and evaluation, determine how environmental stressors in any particular context are perceived and assimilated.

Whether reactions to stressors are eustressful or distressful may be partially dependent on these psychological processes.

For example, Epstein's (1991) cognitive-experiential self theory posits that the self has beliefs about the world. A benign set of beliefs would view the world as a source of pleasure, events as meaningful, people as desirable to relate to, and the self as a positive entity. Epstein reports that when wartime combat leads to the erosion of such beliefs, post-traumatic stress disorder is more frequent. Those soldiers who experienced such decline had unfavorable experiences with respect to the "duration of exposure, leadership and uncertainty."

Whereas the cognitive-experiential model has thus far been used to explain the transposability of negative events to other contexts by way of abstract belief systems, such approaches need to consider how negative experiences can strengthen individuals as well. Elder and Clipp (1988) show that World War II veterans who experienced heavy combat possessed better coping and managerial skills in 1985 when compared to other veterans. Their observation raises the difficult issue of when primary distress leads to secondary eustress. Perhaps highly dangerous episodes that are endured with some measure of success foster unassailable beliefs that one is invulnerable, capable, or "lucky." In the case of broad historical events like war, how the larger society constructs and maintains such events in collective memory can facilitate this conclusion by providing symbolic images of heroism, duty, and honor (Burke 1989); of course, these representative processes can also work against the individual (e.g., veterans of Vietnam).

Summary

Secondary eustress derives from added capacity developed in prior eustressful circumstances. There are at least three major mechanisms of secondary eustress: (1) thresholds of reactivity are raised, (2) motivational structures are enhanced, and (3) individuals select potentially more challenging, but controllable, contexts. (Raised thresholds of reactivity and selection into controllable environments result in the avoidance of potentially distressful stressors, but this avoidance can produce eustressful results because the actor views his or her self as efficacious.) Social psychological mechanisms which explain the secondary eustress processes include dynamic models of efficacy and motivation, as well as learning-generalization. Models proposed by White and Atkinson suggest that individuals who experience greater success in early encounters with stressors have higher levels of motivation with which to respond to later, potentially stressful encounters. Raynor's research suggests that future orientations are of consequence as well: when a stressor is seen as part of a contingent path (i.e., without its resolution, a valued future state is foreclosed), the coping response should be more effective. Kohn's learning-generalization mechanism also can account for these dynamics, with particular emphasis on why individual select into situations which offer the possibility of primary eustress.

DISCUSSION

While an emphasis on the negative consequences of stressors is warranted because stressors challenge the individual's psychological and physical resources, research must also consider the possibility of positive outcomes. Indeed, in a biological context Selye argued that distress is a relatively rare response to stressors. Moreover, at least certain types of life events are more frequently successfully, rather than unsuccessfully, resolved (Thoits 1994; Turner and Avison 1992). Clearly, there is a need to apply a wide range of social psychological theories to these phenomena. Our paper is intended as a preliminary effort in this direction.

At the same time, as a largely deductive exercise, the proposed model has limitations. The postulated relationships are presumed to be linear, although this is almost certainly not the case. For example, one social psychological mechanism which explains secondary eustress is the enhancement of the self-system; the threshold of reactivity to potential stressors is raised and fewer phenomena initiate the alarm phase. Yet, as recognized by psychoanalytic theory (Brenner 1963), stress can serve an important signal function which prompts the actor to deal with contextually-based threats. At some point, high levels of reactivity to stressors become dysfunctional. Similarly, as suggested by the Yerkes-Dobson Law, individuals who are too highly motivated may be among the least capable copers. These examples suggest that these largely deductive propositions may take complicated, non-linear empirical forms (Leik 1979).

There are also macro dimensions to the stress process which are not elaborated in the model. Stressors and coping strategies are not randomly distributed but vary by social location. One's position in virtually any differentiated social system (an organization, a small group, a society) presents a set of potential stressors and at the same time constrains coping options. Some coping behaviors are more efficacious than others. Ideally, a structural position's set of commonly experienced stressors coincides with a set of effective coping options. However, common experience indicates that this is not always the case. The elaboration of macro/micro linkages in the stress process—centering on systemic stressors, the self, and mechanisms of adaptation such as learning-generalization, motivation, and selection—is needed.

Juxtaposing eustress and distress and their disparate implications for the self raises questions about the determinants of each respective sequence. Are the stressors which promote eustress and distress of fundamentally different character? It might be thought that the distinctive features of the stressors themselves are the most critical factors—each having uniformly salutary or deleterious consequences. However, the perspectives reviewed here alert us to differences in individuals' responses to the same stressful situations depending on their life-histories. The proposed model suggests that longitudinal data are necessary to study the dynamics of stress, data which allow for a detailed study of past stressful experiences—including their resolution—and the social psychological processes by which meaning is constructed in the present.

ACKNOWLEDGMENTS

The authors wish to thank Robin Simon and anonymous reviewers for helpful comments. An earlier version of this paper was presented at the 1989 Midwest Sociological Society Meetings, St. Louis. Address all correspondence to the first author: Department of Human Development and Family Studies, The Pennsylvania State University, 110 Henderson Building South, University Park, PA 16802-6504.

NOTES

1. However, analogical transfers from early biological conceptions of tension, homeostasis, and adaptation to developmental psychology were very common and productive. The idea that stress fosters development can be found in most developmental psychology, especially since Piaget used the concept of equilibration. Piaget (1971) adopted the equilibration terminology as an explanatory principle in development following traditions in the biological sciences (Bailey 1984). And, as Langer (1969) notes, psychological theories of development almost unanimously include equilibration, the major exception being behaviorist models. For example, psychoanalytic approaches emphasize the balancing of tensions among the id, ego, and super-ego (Silberman 1981).

2. The proposed model only addresses the maintenance or enhancement of the self as motivational, although other theories of the self postulate that congruency or consistency is a central dynamic (see Gecas and Burke 1995).

3. Schwalbe (1986) provides further discussion of the aesthetic experience.

4. Related concepts are referenced in White's (1959) essay on effectance motivation, especially his discussion of optimal excitation and optimal stimulus (pp. 313-315).

5. Atkinson's model, and its subsequent elaborations, actually suggests that individuals will engage in coping efforts only when their motive to approach success is greater than their motive to avoid failure. For simplicity's sake, we discuss individuals whose motive to approach success exceeds their motive to avoid failure.

6. Some of this research was influenced by the Yerkes-Dobson Law, which posits an inverted U-shaped curve describing the relation between motivation and efficiency: extremely motivated individuals, as well as extremely unmotivated individuals, are likely to be the least efficient actors. Thus, the most highly motivated individuals may actually be poor copers.

7. Social learning theory broadens this perspective considerably: processes of imitation and vicarious learning link the stressful experiences and coping strategies of networks of individuals (Bandura and Walters 1963). Whether by direct learning or socially-based learning, individuals can learn from both distressful and eustressful experiences.

REFERENCES

Aneshensel, C. S. 1992. "Social Stress: Theory and Research." Pp. 15-38 in *Annual Review of Sociology,* Vol. 18, edited by J. Blake and J. Hagan. Palo Alto, CA: Annual Reviews, Inc.

Atkinson, J. W. 1964. *An Introduction to Motivation.* Princeton, NJ: Van Nostrand Reinhold.

Atkinson, J. W., and J. Raynor. 1974. *Motivation and Achievement.* Washington, DC: V. H. Winston.

Bailey, K. D. 1984. "Equilibrium, Entropy, and Homeostasis: A Multidisciplinary Legacy." *Systems Research* 1: 25-43.

Bandura, A. 1993. "Perceived Self-Efficacy in Cognitive Development and Functioning." *Educational Psychologist* 28: 117-148.

Bandura, A., and R. H. Walters. 1963. *Social Learning and Personality Development.* New York: Rinehart and Winston.

Bem, D. J. 1972. "Self-Perception Theory." Pp. 2-62 in *Advances in Experimental Social Psychology,* Vol. 6, edited by L. Berkowtiz. New York: Academic Press.

Breer, P. E., and E. A. Locke. 1965. *Task Experience as a Source of Attitudes.* Homewood, IL: Dorsey.

Brenner, C. J. 1963. *An Elementary Textbook of Psychoanalysis.* New York: International Universities Press.

Burke, K. 1989. *On Symbols and Society,* edited by J. Gusfield. Chicago: University of Chicago Press.

Caplan, L. J., and C. Schooler. 1992. "The Role of Analogy and Processing Complexity in Memory and Inference for Text." Unpublished manuscript.

DeCharms, R., and M. S. Muir. 1978. "Motivation: Social Approaches." Pp. 91-113 in *Annual Review of Psychology,* Vol. 29, edited by M. Rosenzweig and L. Porter. Palo Alto, CA: Annual Review, Inc.

Csikszentmihalyi, M. 1990. *Flow: The Psychology of Optimal Experience.* New York: Harper and Row.

Dohrenwend, B. P., B. G. Link, R. Kern, P. E. Shrout, and J. Markowitz. 1990. "Measuring Life Events: The Problem of Variability Within Event Categories." *Stress Medicine* 6: 179-187.

Edwards, J. R., and C. L. Cooper. 1988. "The Impacts of Positive Psychological States on Physical Health: A Review and Theoretical Framework." *Social Science and Medicine* 27: 1447-1459.

Elder, G. H., Jr., and E. C. Clipp. 1988. "Combat Experience, Comradeship, and Psychological Health." Pp. 131-156 in *Human Adaptation to Extreme Stress: From the Holocaust to Vietnam,* edited by J. P. Wilson, Z. Harel, and B. Kahana. New York: Plenum Press.

Elder, G. H., Jr., L. George, and M. J. Shanahan. 1996. "Psychosocial Stress Over the Life Course." In *Psychosocial Stress,* edited by H. Kaplan. San Diego: Academic Press.

Entin, E. E., and J. O. Raynor. 1973. "Effects of Contingent Future Orientation and Achievement Motivation on Performance in Two Kinds of Tasks." *Journal of Experimental Research in Personality* 6: 314-320.

Epstein, S. 1991. "Cognitive-Experiential Self Theory: Implications for Developmental Psychology." Pp. 79-123 in *Self Processes and Development: The Minnesota Symposia on Child Psychology,* Vol. 23, edited by M. R. Gunnar and L. A. Sroufe. Hillsdale, NJ: Lawrence Erlbaum Associates.

Erickson, R. J. 1993. "The Importance of Authenticity for Interrelating Society, Self, and Emotions." Paper presented at the Annual Meeting of the American Sociological Association, Miami, FL.

Gecas, V. 1986. "The Motivational Significance of Self-Concept for Socialization Theory." Pp. 131-156 in *Advances in Group Processes,* Vol. 3, edited by E. Lawler. Greenwich, CT: JAI Press.

Gecas, V., and P. J. Burke. 1995. "Self and Identity." Pp. 41-67 in *Sociological Perspectives on Social Psychology,* edited by K. S. Cook, G. A. Fine, and J. S. House. Boston: Allyn and Bacon.

Hall, C. S., and G. Lindzey. 1957. *Theories of Personality.* New York: John Wiley.

Heider, F. 1958. *The Psychology of Interpersonal Relations.* New York: Wiley.

Karasek, R. A., R. S. Russell, and T. Theorell. 1982. "Physiology of Stress and Regeneration in Job Related Cardiovascular Illness." *Journal of Human Stress* 8: 29-42.

Kiecolt, K. J. 1994. "Stress and the Decision to Change Oneself: A Theoretical Model." *Social Psychology Quarterly* 57: 49-63.

Kohn, M. L., and C. Schooler, with the collaboration of J. Miller, K. A. Miller, C. Schoenbach, and R. Schoenberg. 1983. *Work and Personality: An Inquiry Into the Impact of Social Stratification.* Norwood, NJ: Ablex Publishing Corporation.

Langer, J. 1969. "Disequilibrium as a Source of Development." Pp. 22-37 in *Trends and Issues in Developmental Psychology,* edited by P. Mussen, J. Langer, and M. Covington. New York: Holt, Rhinehart and Winston.

Lazarus, R. S. 1991. *Emotion and Adaptation*. New York: Oxford
Leik, R. K. 1979. "Let's Work Inductively, Too." Pp. 177-191 in *Contemporary Issues in Theory and Research*, edited by W. Snizek, E. Fuhrman, and M. Miller. Westport, CT: Greenwood.
Lott, B., and A. J. Lott. 1985. "Learning Theory in Contemporary Social Psychology." Pp. 109-135 in *Handbook of Social Psychology*, Vol. I, edited by G. Lindzey and E. Aronson. New York: Random House.
Markus, H., and E. Wurf. 1987. "The Dynamic Self-Concept: A Social Psychological Perspective." Pp. 299-337 in *Annual Review of Psychology*, Vol. 38, edited by M. Rosenzweig and L. Porter. Palo Alto, CA: Annual Reviews, Inc.
Mead, G. H. 1926. "The Nature of the Aesthetic Experience." *International Journal of Ethics* 36: 382-393.
_____. 1938. *The Philosophy of the Act*, edited by C. Morris. Chicago: University of Chicago Press.
Menaghan, E. G., and E. S. Merves. 1984. "Coping With Occupational Problems: The Limits of Individual Efforts." *Journal of Health and Social Behavior* 25: 406-423.
Miller, J. 1988. "Jobs and Work." Pp. 327-359 in *Handbook of Sociology*, edited by N. J. Smelser. Beverly Hills, CA: Sage.
Mirowsky, J., and C. E. Ross. 1989. *Social Causes of Psychological Distress*. New York: Aldine de Gruyter.
Mortimer, J. T., and R. G. Simmons. 1978. "Adult Socialization." Pp. 421-454 in *Annual Review of Sociology*, Vol. 4, edited by R. Turner, J. Coleman, and R. Fox. Palo Alto, CA: Annual Reviews, Inc.
Piaget, J. 1971. *Biology and Knowledge*. Chicago: University of Chicago Press.
Raynor, J. O. 1969. "Future Orientation and Motivation of Immediate Activity: An Elaboration of the Theory of Achievement Motivation." *Psychological Review* 76: 606-610.
Ross, M., and G. J. O. Fletcher. 1985. "Attribution and Social Perception." Pp. 73-122 in *Handbook of Social Psychology*, Vol. II, edited by G. Lindzey and E. Aronson. New York: Random House.
Schneider, D. J. 1991. "Social Cognition." Pp. 527-561 in *Annual Review of Psychology*, Vol. 42, edited by M. Rosenzweig and L. Porter. Palo Alto, CA: Annual Reviews, Inc.
Schooler, C. 1987. "Psychological Effects of Complex Environments During the Life Span: A Review and Theory." Pp. 24-49 in *Cognitive Functioning and Social Structure Over the Life Course*, edited by C. Schooler and K. W. Schaie. Norwood, NJ: Ablex.
Schwalbe, M. L. 1986. *The Psychosocial Consequences of Natural and Alienated Labor*. Albany, NY: State University of New York Press.
Selye, H. 1975. "Confusion and Controversy in the Stress Field." *Journal of Human Stress* (June): 37-44.
_____. 1976. *The Stress of Life*. New York: McGraw Hill.
Silberman, I. 1981. "Balance and Anxiety." *Psychoanalytic Study of Children* 36: 365-380.
Smith, W. H. 1987. "The Stress Analogy." *Schizophrenia Bulletin* 13: 215-220.
Thoits, P. A. 1994. "Stressors and Problem-Solving: The Individual as Psychological Activist." *Journal of Health and Social Behavior* 35: 143-159.
Thorndike, E. L. 1913. *The Psychology of Learning*. New York: Teachers College, Columbia University.
Turner, R., and V. Billings. 1991. "The Social Contexts of Self-Feeling." Pp. 103-122 in *The Self-Society Dynamic: Cognition, Emotion, and Action*, edited by J. A. Howard and P. L. Callero. New York: Cambridge University Press.
Turner, R. J., and W. R. Avison. 1992. "Innovations in the Measurement of Life Stress: Crisis Theory and the Significance of Event Resolution." *Journal of Health and Social Behavior* 33: 36-50.
Turner, R. J., B. Wheaton, and D. A. Lloyd. 1995. "The Epidemiology of Social Stress." *American Sociological Review* 60: 104-125.
Wheaton, B. 1980. "The Sociogenesis of Psychological Disorder: An Attributional Theory." *Journal of Health and Social Behavior* 21: 100-124.
_____. 1983. "Stress, Personal Coping Resources, and Psychiatric Symptoms: An Investigation of Interactive Models." *Journal of Health and Human Behavior* 24: 208-229.

_____. 1990. "Life Transition, Role Histories, and Mental Health." *American Sociological Review* 55: 209-223.
_____. 1994. "The Domains and Boundaries of Stress Concepts. Paper presented at the 1994 Meetings of the American Sociological Association, Los Angeles, CA.
White, R. H. 1959. "Motivation Reconsidered: The Concept of Competence." *Psychological Review* 66: 297-333.
Zimmerman, R. S. 1988. "Social Psychology and Health: Psychological Social Psychology's Contributions." Paper presented at the American Sociological Association Meetings, Atlanta, GA.

THE FAMILY DEVALUED:
THE TREATMENT OF THE FAMILY IN SMALL GROUPS LITERATURE

Lala Carr Steelman and Brian Powell

ABSTRACT

Although the family is widely appreciated as an influential small group, rarely do small group researchers apply their theoretical advances to it. We review one specific area in familial research that might spur interest among small groups researchers—the link between the structure of the sibling group and educational outcomes. Sibship configuration specifically refers to features such as size, sex composition, birth order, and age spacing. We review and synthesize previous evidence on the influence of these four parameters of sibship, discuss ongoing scholarship, and recommend future directions in this area. It is notable that size and sex composition have commanded attention in both family and small group research. Nonetheless, despite many seeming commonalities, an invisible wall still seems to partition family researchers from small group scholars. Reasons for this division are discussed, and some ways to bridge the gap are advanced.

Advances in Group Processes, Volume 13, pages 211-236.
ISBN: 0-7623-0005-1

INTRODUCTION

Virtually all social scientists acknowledge that the family is a pivotal small group. It is standard fare in introductory sociology and social psychology textbooks to provide the family as the paradigmatic example of a primary group—one that is small, intimate, enduring, and profoundly influential over its members (Cooley 1909). Curiously, however, the small groups literature expends less energy on the family than it may well warrant. Indeed, in this very series of manuscripts devoted to recent breakthroughs in our knowledge of small groups, it is notable that over the past decade only a handful of articles has dealt with the family either directly or tangentially (for an exception, see Grusky, Bonacich, and Peyrot 1988). Recent books on group processes also fail to set the family apart as a worthy group to explore (Foschi and Lawler 1994). A fresh look at the family literature, specifically that which addresses structural parameters as a potential contributor toward understanding group dynamics, thus seems like a wise endeavor. Should a joint body of findings emanate from the study of small groups and the study of one specific small group, the family, then some general principles of group dynamics might materialize.

Because the family stands at the crossroads between the individual and the society-at-large, it has surfaced as a strategic group to study that cuts deeply across disciplinary boundaries. One prominent area in which the small groups literature and the family may dovetail is the implications of structural features of the family, such as parental structure (e.g., the differences among one-parent, two-parent, and step-parent households) and sibship composition (such as number of siblings). While the study of structural parameters of the family may seem narrow to some, there may be more general applicability than first meets the eye. Social psychologists who devote their scholarly attention to how small groups behave have much to share with those concerned with familial dynamics. Both strands of inquiry are curious about effects of structural characteristics of the group, such as group size, member ranking, and sex composition, on the individual behavior of members and on collective outcomes. Regrettably, few scholars working in the small groups tradition or, we should add, writing on the sociology of the family envision how greater insight into family structure may ultimately illuminate the study of small groups. Somehow this rather obvious association has not been made. Parenthetically, while the small groups literature has been enormously successful in developing rich theories, it has been limited by its reliance on data collected almost exclusively on unacquainted laboratory subjects. By investigating the consequences of group structure on two fronts, the more theoretical tests of group processes performed in laboratory settings and the perhaps less structured probes into the family, rich empirical and theoretical dividends may ultimately be realized.

Of course, for any cross-fertilization of these ideas to occur, there must be communication between family and small groups researchers. To facilitate this goal, in this paper we present some of the issues addressed in family research which may be of interest to those studying small groups. We focus specifically on one set of structural parameters in the family—namely, sibship configuration—and its educational consequences. In particular, we report on four components of sibship structure:

1. Size of the sibling group—the number of children in the family.
2. Sex composition—the relative number of girls vs. boys in the sibship.
3. Birth order—a child's position in the age hierarchy of siblings in the family.
4. Child spacing—the time intervals that separate the births of children.

Of these structural characteristics, sibship size and sex composition have direct parallels in the literature on small groups (Steelman 1985), whereas birth order and child spacing do not unless one stretches the definitions a bit. For example, ordinal position and child spacing may tap into the status characteristic of age.

We hope to make the case for greater dialogue between seemingly different groups of researchers—small groups and family sociologists—who actually may have much in common. We review and synthesize previous evidence on the influence of these four sibship parameters and discuss on-going work in this area. After covering the ground where we have been, our discussion then turns to where we should now go in this line of inquiry. We conclude by posing two questions which cannot be fully answered: (1) Why has there been an invisible wall between small groups and family research, and (2) are there mutual areas in which both genres of research can exchange ideas?

THE CURRENT STATUS OF FAMILIAL STRUCTURE RESEARCH

Given increasing variance of family forms in the United States, it is no wonder that the implications of family structure have assumed a more visible place in scholarly debate than they previously enjoyed. Concern continues to build over the ramifications of factors such as marital disruption, labor force participation of parents, out-of-wedlock births, declining birthrates, and what these conditions imply for the American family in addition to those in other countries experiencing similar trends (Amato, Loomis, and Booth 1995; Bumpass 1984; Downey and Powell 1993; Furstenberg and Cherlin 1991; Hetherington 1989; McLanahan and Booth 1989).

Because of the high profile of these issues, many scholars assume that family structure refers chiefly to parents and their marital histories or to parent/child interactions. Our discussion in this paper, however, centers on the structural configuration of the sibling group and its potential impact on educational outcomes for children. Admittedly a narrow focus, this area is one in which family research has evolved demonstrably, made tremendous strides, and deserves to be considered alongside sophisticated analyses of small groups that are currently being generated.

While research on the impact of family structure has taken many fascinating twists and turns, such as efforts to link birth order to just about every conceivable outcome from stripteasing to reactions to the San Francisco earthquake, the most impressive work has been done on the relationship between sibling group features and educational outcomes. The literature in this area has grown from making simple observations about the relationships among family size, birth order, and educational attainment to formalizing and testing acclaimed and controversial theories to account for these patterns (Blake 1989; Downey 1995; Gailbraith 1982; Heer 1985, 1986; Lindert 1974; Retherford and Sewell 1991; Steelman and Powell 1991; Sulloway 1995, in preparation; Teachman 1987; Zajonc and Markus 1975).

Although it has failed to capture the imagination of those in the area of small groups, the issue of structural constraints of the family merges the concerns of many specialties within the field of sociology. Long fascinated with the antecedents of fertility behavior, demographers have become increasingly intrigued by its aftermath (Blake 1981, 1986, 1989; Bumpass, Rindfuss, and Janosik 1978; Goldscheider and Goldscheider 1989). Moreover, since the family may provide the mechanism by which the greater society transmits its influence to members, it also has gained substantial notice from status attainment researchers. After all, Blau and Duncan's (1967) classic investigation of status attainment incorporated the number of siblings as a predictor of life chances. Others have followed quite admirably in their footsteps and sibship size has become a standard inclusion in status attainment models (Alwin and Thornton 1984; Featherman and Hauser 1978; Retherford and Sewell 1991; Sewell, Hauser, and Wolf 1980).

Outside sociology's realm, there also stands an influential body of literature on the nexus between sibling configuration and a wide array of outcomes. Economists, particularly those who embrace the human capital model of achievement, are interested in the link between family structure and resource allocation, especially with respect to monetary exchanges between parents and children, to paying for college and/or to apportioning inheritance among surviving offspring (Becker 1964, 1981; Becker and Tomes 1976; Taubman and Behrman 1986). Evolutionary biologists have attempted to specify the conditions under which sibship structure may motivate parental investments to maximize genetic fitness (Hdry 1987). Psychologists have concentrated on other connections to family structure, including the intellectual atmosphere to which children are exposed as a function of sibship constellation (Zajonc and Markus 1975) and the interplay among ordinal position, other aspects of the sibship structure, and sibling rivalry, role-taking behaviors, feelings of anxiety, and orientations to success. Even some family therapists are drawn to research in this area, although their interests usually emphasize social and interactional, rather than educational, domains. Adler's (1959) work on birth order, for example, convinced some psychologists and clinicians that information regarding birth order was useful in devising intervention strategies and helping with marriage counseling. Given that the study of the family has widespread visibility across a number of fields and that the family represents a pivotal small group that thrives in the natural setting, we would expect small groups researchers to search for common themes with their own agenda.

Size of the Sibling Group

Historically, interest in the impact of family configuration on children's intellectual development probably began with Sir Francis Galton's *English Men of Science* (1874) in which he pronounced the "superiority of the eldestborn child." According to Galton, first-born children were more likely than their later-born counterparts to be counted among the most eminent scientists. Despite the appeal of this argument and subsequent, often anecdotal, accounts on the power of birth order, however, scholarly preoccupation with ordinal position has given way to more focused attention to family size. This shift probably occurred because many of the pioneer efforts in examining the relationship between birth order and intellectual development failed to recognize the correlation between birth order and the size of the sibling group. Put simply, raw comparisons were being made between people from varying birth orders without regard to the size of the family from which they originated. Thus, fifth-born children, who by definition come from large families, were being compared to first-born children, who could come from any family size but most likely came from much smaller ones. Indeed, the correlation between family size and birth order is quite high (around 0.65). Therefore, failure to consider family size led many of the earlier researchers to overstate the impact of birth order, especially with respect to intellectual development.

Over a decade ago, Steelman and Mercy (1980) and Page and Grandon (1979) independently arrived at the same conclusion that, once family size is taken into account, birth order is unrelated to intellectual development. Those papers initially encountered much resistance in the intellectual community in part because they contradicted conventional wisdom. Nevertheless, their observations withstood intense scrutiny. Most contemporary researchers concur that family size is strongly and inversely related to educational performance while birth order is not (Blake 1989; Hauser and Sewell 1985; Retherford and Sewell 1991). Amid these developments, many researchers reoriented their focus toward trying to figure out the influence of family size rather than clinging to the belief that birth order altered intellectual development and educational performance (Ernst and Angst 1983).

Even prior to this shift, status attainment researchers had long incorporated the number of siblings into their models and had observed a negative influence on educational and other forms of attainment. This relationship persisted across various measures of academic success, social class conditions, cultures, and other familial permutations. Rare were deviations from this pattern. Among the exceptions, Rankin, Gaite, and Heiry (1979) failed to replicate the impact of family size

for American Samoan children, as did Gailbraith (1982) and Gomes (1984) in
their analysis of Mormon college students and Kenyans, respectively. Irrespective
of these infrequent departures, which have been critiqued heavily by Blake (1989),
the inverse relationship between family size and educational outcomes, be they
educational performance (such as grades in school or standardized test performance) or educational attainment (such as high school graduation or college
matriculation), is one of the most firmly grounded observations in the social science literature (Cicirelli 1978; Nisbet and Entwistle 1967; Steelman 1985). Perhaps as a by-product, practically all of the recent large data sets used by social
scientists to study educational and other status attainment outcomes—including
the National Longitudinal Survey of the High School Class of 1972, High School
and Beyond, the National Educational Longitudinal Study of 1988, the National
Survey of Families and Households, and the General Social Survey—now provide
information on sibship size. And, in a somewhat circular fashion, in part because
of its convenient access in these data sets, the size of the sibling group is the most
consistently evaluated structural characteristic of the sibship.

Not surprisingly, the effect of sibship size resembles that of socioeconomic status because both variables constrain the amount of resources that can be distributed to any child. Socioeconomic status taps the pool of resources to some extent
while the number of people among whom the resources are to be divided is gauged
by the number of children in the family. Yet, while many sociologists underscore
the influence of socioeconomic status, fewer focus on the impact of the number of
siblings. Instead, despite its nearly unfailing relationship with status outcomes,
sibship size is often relegated to the role of control variable or afterthought.
Although some social scientists wrangle over whether the size of the sibling group
renders its greatest impact in the early or later years of childhood (Alwin and
Thornton 1984; Blake 1986, 1989), there is agreement that the effect is probably
not artifactual. Beyond this observation, many researchers are content to recognize the existence of the pattern without digging deeper to uncover its theoretical
meaning.

One can ferret out six basic explanations that have been advanced in the literature to account for the impact of the size of sibling group. The first four that we
cover here cannot be summarily dismissed; however, they enjoy less advocacy
than the last two.

Confounding Variables

First, sibship size may be related to other confounding variables that also sway
educational advancement. The possibility of spurious correlation, thus, naturally
arises. To illustrate, socioeconomic status generally is related negatively to fertility and positively to educational success. Seeming effects of sibship size may,
therefore, be an artifact of socioeconomic differentials in childbearing.

Although at first glance this explanation appears plausible, it can be fairly easily discounted. In most studies, once controls for socioeconomic status are included, the impact of the size of the sibling group, while reduced, ordinarily persists. For example, Powell and Steelman's (1993) recent analysis of the High School and Beyond data set indicated that only 25 percent of the positive relationship between sibship size and high school attrition is attributable to socioeconomic factors— specifically, parental education, parental structure (i.e., two- vs. single-parent households), familial income, and race. For college graduation (to which sibship size is negatively linked), the figure increases slightly to approximately 30 percent. In terms of educational performance, one-third of the negative effect of sibship size on standardized test scores is accounted for by these factors. And while certainly not all potentially confounding variables have been explored, other researchers who have incorporated a spate of possibly contaminating factors associated with family size into their models are impressed with the resiliency of this size variable (Blake 1989; Downey 1995).

Physiological/Genetic Weaknesses

Second, the negative association between sibship size and educational outcomes may be a sign of either physiological or genetic weaknesses (Broman, Nichols, and Kennedy 1975; Grotevant, Scarr, and Weinberg 1977). The physiological argument is that women who continue bearing children expose their progeny to at-birth risks such as low birth weight and make them less physically resilient. In turn, these medical traumas depress academic performance. This explanation implies birth order effects in addition to family size ones. According to the genetic argument, people who choose to have large families are those who possess less genetic talent and, in turn, pass this deficiency onto their children.

Evidence to buttress either of these two views is slim (Powell and Steelman 1993). Higgins, Reed, and Reed (1962), for example, argued that the greater vulnerability of verbal skills (as opposed to nonverbal skills) to environmental input undermines the genetic argument. In addition, Mercy and Steelman (1982) suggested that the uneven influence of maternal and paternal education further weakens the genetic explanation since both parents contribute equally to genetic makeup. Finally, using the reasoning of the explanation, one should expect the effect of sibship size on educational attainment to be entirely the result of the link to educational performance (i.e., grades and test scores). Powell and Steelman (1993), however, found that approximately two-thirds of the effect of sibship size on both high school graduation and college entrance persists even upon including grades and standardized tests scores in their models.

Parental Fertility Decisions

Third, parents who place a lower premium on education also may prefer and choose to have larger families. As some human capitalists might capture it (Becker and Tomes 1976), "quantity" then is selected over "quality" of children. However, Blake (1989) convincingly identified various problems with this logic. This reasoning, for example, assumes that fertility decisions are purely, or even primarily, rational. It also presupposes that decisions regarding anticipated educational investments are more salient than other factors in determining fertility patterns— an unlikely proposition. Most social scientists, therefore, conclude that the consequences of family structure are more influential in educational outcomes than are events preceding family structure. The question of causality, nevertheless, still looms large and has not been completely answered. Ultimately resolving this conundrum will require longitudinal data.

Socialization Explanations

Fourth, sibship size may influence socialization practices which, in turn, shape educational outcomes. In contrast to the other explanations, this one represents a hybrid of several diffuse ideas regarding sibship size—which were suggested several decades ago but, ironically, are not typically used to examine the specific relationship between sibship configuration and educational progress.

Among this set of ideas is that sibship size may be linked to differential valuation of obedience and self-direction, in line with the influential work of Kohn and associates (Kohn 1969; Kohn, Slomczynski, and Schoenback 1986). The very nature of large families may dictate greater emphasis on obedience. As the number of children in the household increases, the combinations of interactive groups increase exponentially. Thus, growing family sizes may require more explicit rules, more enforcement of rules, and more bureaucratization in daily familial administration—resulting in a more autocratic style of parenting and a higher premium placed on obedience than on independence (Bossard and Boll 1956; Elder 1985; Elder and Bowerman 1963; Nye, Carlson, and Garrett 1970). Further, the heightened interdependence among its members may prompt parents in a large family to reward cooperation and downplay competition (Rosen 1961). Other social psychologists focus on how large families may insulate their members from interaction with those beyond their boundaries. Such isolation may restrict children's knowledge of role behavior and decrease their ability to take the role of the other (Heise and Roberts 1970). If self-direction, competition, and role-taking ability are needed for the development of intellectual skills and school performance, one should expect a negative relationship between size of sibling group and educational outcomes.

While these suggestions are tantalizing, they have not been persuasively documented in empirical work. For the most part, they simply have not been capitalized upon in contemporary scholarship by family and small groups researchers. Yet these concepts are more explicitly rooted in small group theory than many other explanations of the sibship/education link. When these ideas have been evaluated, however, research has not offered unequivocal support. Some scholars, for example, question whether the link between sibship size and parental value orientations remains once appropriate socioeconomic controls are taken into account (Alwin 1984).

One can certainly see how even a simple bivariate association such as that between the size of the sibling group and educational outcomes can become quite complicated. In the myriad of explanations advanced, the next two theoretical models stand out as the most innovative and most widely accepted thinking on the topic.

Confluence Theory

The fifth explanation, the confluence theory, was first proposed by Zajonc and Markus (1975) to explicate the relationships thought to exist among family size, birth order and IQ. According to the confluence model, which is the theory's mathematical expression, intellectual development is seen as a function of the intellectual environment to which the developing child is exposed. Intellectual atmosphere is operationalized as "some function of the absolute intellectual level of its (the family's) members" (Zajonc and Markus 1975, p. 227). Zajonc and Markus' definition of intellectual level is not identical with IQ. In contrast to the latter, intellectual level is not adjusted for age differences. Rather, since children are not as intellectually mature as adults, their presence in the family dampens its intellectual ambience. Zajonc and Markus contended that camaraderie with intellectual superiors will sharpen individuals' intellectual acumen. Correspondingly, the greater the number of young children, the greater deterioration in the intellectual climate of the household. Because the average intellectual environment declines with each successive birth, later-born children generally should exhibit less intellectual talent than their older siblings who had the advantage of growing up (for however brief a period) without the intellectual diminution supplied by their younger siblings. The relative disadvantage of being a later-born or having many siblings will depend in part on how closely spaced in age siblings are. This theory also implies that, in the case of twins or very closely spaced siblings, one would expect a depressed intellectual environment.

One anomaly requiring additional clarification is that only children do not outperform other children in small sibships. The confluence theory, as posited above, suggests that only children should reap the greatest academic benefit because there are no other children to dilute the academic atmosphere. Yet most data do not fit these expectations. As a result, Zajonc and Markus added an extra component to their theory: a teaching function in which those who are in the position to teach younger siblings are intellectually stimulated whereas those who are not (i.e., singletons and last-borns) suffer a teaching deficit or handicap.

One sign of a promising theory or framework is that it inspires research. The confluence theory more than fulfills this requirement as a profusion of studies has appraised its tenets. In fact, this provocative and initially much lauded theory has elicited heated reactions by both advocates and critics. However, the weight of evidence seems to side with the latter group. Many scholars have questioned the fit between this creative theory and the data that have been analyzed to assess it (Blake 1989; Ernst and Angst 1983; Hauser and Sewell 1985; Olneck and Bills 1979; Retherford and Sewell 1991; Steelman and Mercy 1980). Some conclude that while the impact of family size endures, other implications of the model perform much more poorly (Steelman 1985). For example, that the patterns regarding birth order do not square with the model's predictions undermines its credibility (Blake 1989; Hauser and Sewell 1985).

Even if these criticisms lack merit, the confluence theory, similar to the physiological/genetic weakness explanation, is not entirely successful in untangling the connection between sibship configuration and educational attainment. As mentioned earlier, the link between sibship size and high school graduation and college attendance is not solely a function of the relationship to intellectual ability. Rather, this link remains even upon taking into consideration grades and standardized test performance. Some scholars have looked toward other possibilities to explain the pattern. One of the more alluring alternatives is the resource dilution hypothesis to which we now turn.

Resource Dilution Model

With inconsistent support for the confluence model, scholars have offered another deceptively simple account of the relationship between family size and educational outcomes. This sixth explanation, the resource dilution model, is probably the most palatable to sociologists (Blake 1989; Steelman and Powell 1989); it also enjoys considerable support among economists (Becker 1964, 1981; Taubman and Behrman 1986). Stated simply, the resource dilution argument holds that, as the number of members in the family increases, the ability of the family to provide resources, whether these be intellectual, cultural, social, or economic, declines correspondingly (Anastasi 1956). This dilution of resources has ramifications for intellectual development and educational progress. In other words, the more children in the family, the fewer intellectually lucrative resources that can be provided to any given child (Powell and Steelman 1990). Whereas the confluence and physiological/genetic weakness explanations are unable to put forth a reason for the effect of sibship size on educational attainment net of intellectual ability, the resource dilution argument more easily absorbs this pattern. It holds that while some resources may influence academic ability (and via academic ability indirectly affect educational attainment), other resources, such as financial outlays for college, will affect educational attainment directly.

Blake (1989) suggested that there are three areas in which the number of siblings depletes the finite resources available to them. They are: (1) types of homes, necessities of life, and cultural objects; (2) personal attention, intervention, and teaching; and (3) specific chances to engage the outside world. These clusters parallel the emphases in the literature on the role that economic capital (Becker 1964, 1981), social capital (Coleman 1988) and cultural capital (Bourdieu 1977; DiMaggio 1982) play on the development of youths. While confluence theorists emphasize intellectual stimulation and others stress cultural and social capital, Steelman and Powell (1989, 1991) have, along with several economists (Taubman and Behrman 1986), interjected the sheer input of economic resources into thinking on this topic.

Interestingly, the confluence model is subsumable under the resource dilution account except that the latter broadens the scope of resources that can be distributed to children to encompass nonintellectual factors such as social support, cultural activities, and economic backing.

While the resource dilution hypothesis represents an explanation that fits existing material very nicely, in its earliest rendition it was used in an ad hoc fashion to reconcile empirical evidence. This explanation has experienced substantial modification as critics have pointed out its limitations and those in the field have responded accordingly. Initially most researchers merely assumed that the inverse relationship between family size and educational outcomes was mediated through the allocation of resources to children. Nevertheless, few studies actually directly checked the three essential elements of this argument: (1) resources are diluted as a function of sibship size; (2) reduced resources decrease intellectual ability and educational attainment; and (3) the effect of sibship size on educational progress is primarily via the allocation of resources. The empirical inattention to these elements is a serious oversight. Certainly a solid argument could be made that additional siblings actually enhance the resource base in some ways that benefit children. For example, one could argue that having many siblings to play with is a stimulating experience or that having several younger siblings to tutor or older siblings to be tutored by positively affects academic development. Moreover, while resources may be implicated in educational progress and sibship size may be instrumental in reducing resources, it does not necessarily follow that all, or even the preponderance, of sibship size's effect on academic progress is a function of the link to resources.

In addition, many questions regarding the meaning of resources remain. There has been a lack of clarity in defining resources. We need to specify what is meant by resources and to identify the mechanisms through which resources operate. As noted above, there are a wide range of resources: social, cultural, interactional, economic, and intellectual. It would be naive to think, for example, that economic resources behave in the same fashion as, say, intellectual ones. And it would equally unwise to assume that all economic resources follow identical patterns. Not all resources are fungible. Some resources can be shared, while others are not recoverable. Some may respond differently to sibling configuration. A great deal depends upon the type of resources under investigation. However, few researchers have even taken a stab at unraveling what we mean by resources.

Some studies have begun to fill this void. Mercy and Steelman (1982) found an association between the number of siblings and activities in which children engaged. Activities that depressed intellectual performance on standardized exams, such as television viewing, more likely took place in large families. Other activities that bolstered performance, such as reading books, more likely typified small families. Nonetheless, only a few types of interactions were examined. Lindert (1974) and Hill and Stafford (1974) also found that the amount of time parents devoted to their children declined as a function of family size and mediated the negative impact on ability. Parental encouragement also slid with increases in family size (Blake 1981; Marjoribanks 1972). Blake (1989) showed that dance lessons and travel, items which provide a rough proxy for cultural capital, decreased as family size rose. Teachman (1987) additionally observed that the availability of educational objects such as encyclopedias, books, and a place to study decreased with larger family sizes. Steelman and Powell (1989, 1991) found that parental economic investments, specifically funding set aside or actually spent for their children's college education, declined as family size rose and that these economic resources were directly connected to the likelihood of attending and staying in college. They also demonstrated that educational materials, parental discussions with children, and maternal educational aspirations were negatively related with the size of sibling group (Powell and Steelman 1993) and that resources played a crucial role in school performance (i.e., grades and standardized test scores), high school graduation, and post-secondary school attendance.

Downey (1995) has taken the most sweeping approach in this area by looking at the impact of family size on the allocation of economic resources (money saved for college, cultural activities, and classes), household objects (computers and educational objects), and parental interpersonal investments. In addition to replicating the inverse pattern between family size and resources (and educational outcomes) and the positive relation between resources and educational outcomes, Downey demonstrated that essentially all of the adjusted (i.e., controlling for background characteristics) sibship size effect on grades and standardized math test scores, as well as half of the effect on standardized reading test scores, results directly from the dilution of these resources.

Although it now is fairly evident that allocation of many resources is negatively affected by sibship size, the next step is to specify more precisely the functional form of this relationship. Most studies have assumed a linear relationship—that is, a steep and consistent decline in the amount that could be given to any child as the family grew in size. However, careful examination of how resources are allocated may challenge this presumption. As early as in the work of Anastasi (1956), a few scholars recognized that each additional child entering the family may not have the same detrimental effect as the first one.

Simple mathematics illustrates this concept. Suppose, for example, that a parent has $1,000 of discretionary funding which can be allocated to children. If the parent has one child, the child theoretically could receive the entirety of the funds. If the parent has two children and if the parent dispenses resources equally, each child would receive half (i.e., $500). With three children, the figure would be one-third (i.e., $333). Thus, the impact of each additional sibling tapers as sibship size increases. Using this reasoning, one should look beyond the simple linear term and begin investigating the $1/X$ form of the relationship.

Of course, other nonlinear functions might apply as well. For example, there may be a threshold—that is, the family becomes so encumbered at a certain family size that each additional child beyond this threshold has no extra detrimental effect. To date, these possibilities have not been seriously considered with the exception of Downey who demonstrated that linear, $1/x$, and threshold functions apply for different types of resources.

He also found a double dilution process for some resources in which (1) increased family size reduces the availability of certain resources (such as educational objects in the home); and (2) even if resources are available, youths in large families profit less from them than do their counterparts in small families.

In summary, this seemingly straightforward area of research has gone beyond its embryonic stage and has entered a period of accelerated maturation. From the simple observation that family size may deleteriously impact educational outcomes, many noteworthy developments have flourished. Several theoretical models have been elaborated to explicate the impact of sibling structure—some more successfully than others. What is really interesting about this whole line of inquiry is that it orbits around the utterly simple concept of size. To some this variable may seem too basic to whet the intellectual appetite. However, to others trying to understand the implications of this magnificently simple variable may open many doors. If we detach ourselves for a moment and scrutinize other areas of scientific investigation, we also see the relevance of the size of groups as an important determinant of a host of outcomes. Across a wide spectrum of units (such as organizations and populations), size emerges as a powerful variable. Ironically, size of group used to be a focal point of earlier studies of group processes (Bales, Hare, and Borgatta 1957; Caplow 1968; Hare 1952; Simmel 1950; Slater, 1958). And there were fruitful exchanges between small group scholars and those in family studies (Bossard and Boll 1956; Elder and Bowerman 1963; Rosen 1961)—exchanges which unfortunately have waned in frequency in recent years.

Sex Composition

As a rich body of literature on the effects of sibship size has evolved, albeit still requiring elaboration, some scholars have steered their attention to other aspects of sibship constellation—such as sex composition, ordinal position, and age spacing. This redirection has occurred in part because of the increased, although still limited, availability of data on these sibship features. We, however, believe it is more the product of the recognition of demographic changes in the United States and elsewhere. In the United States, for example, the average number of children

224 LALA CARR STEELMAN and BRIAN POWELL

now hovers around two and the variation in family size has been conspicuously reduced (Teachman and Schollaert 1989). Thus, it should come as no surprise that scholars are examining other sibship features in which the degree of variation may be greater.

In a sense, sex composition is a more refined extension of the family size variable partitioned into the number of brothers and the number of sisters (or, alternatively, percentage of sisters/brothers or number of sisters/brothers within each sibship size). Just as size of the group attracts many levels of analysis, so too does sex composition. At the macro level, Guttentag and Secord's seminal work (1983), for example, argued that societal consequences hinge on the sex distribution of the population which can vary dramatically because of wars, infanticide, and gender differences in mortality. At the organizational level, Kanter (1977) skillfully demonstrated the importance of sex composition (especially tokenism) on perceptions and behaviors in corporations. At the micro level, several small groups researchers have debated whether and, if so, how sex composition of groups affects task allocation, perceptions of fairness, attributions, and decision making (Mabry 1985; Ridgeway 1988; Smith-Lovin, Skvoretz, and Hudson 1986). Despite the proliferation of social psychological studies on gender and sex, we do not have a cohesive understanding of the role of sex composition in groups. Indeed, Wiley (1995) categorized findings from this cluster of research (which she merges with research on sex category) as "confusing and often contradictory."

The findings from research on sex composition of the sibship are no less confusing, and there has been a bewildering array of explanations for such observations. Perhaps because the patterns are such a puzzle, scholars have become increasingly intrigued by the possibility that sex composition of sibship influences socialization patterns, parental investments, and educational outcomes.

This topic was broached several decades ago by Koch (1955) and Brim (1958) who posited that those with older opposite-sex siblings exhibit more cross-sex traits (i.e., females displaying "masculine" qualities and males displaying "feminine" ones). If education is deemed a masculine characteristic, as suggested by Becker (1981), youths with brothers should academically outdistance their counterparts with sisters. Of course, one could persuasively challenge the premise that excellence in education is a masculine trait. In fact, because females typically receive higher grades in school than do males, one might reach the opposite conclusion (Powell and Steelman 1990). Furthermore, there is some evidence that the absence of an opposite-sex sibling may create a situation in which cross-sex behavior (e.g., in household tasks) is accepted and encouraged (Brody and Steelman 1985).

Elder and Bowerman (1963) offered a different interpretation and contended that parents with larger families, especially those more heavily populated with boys, more frequently resorted to physical punishment. Downey, Jackson, and Powell (1994) also found that as the proportion of sons increased, mothers were more likely to value obedience in children. While neither Elder and Bowerman nor

Downey, Jackson, and Powell directly tied these findings to educational outcomes, one can imagine such a link—especially if one recalls the socialization explanation for sibship size effects mentioned earlier. That is, a focus on punishment and obedience depresses academic performance; therefore, having brothers should decrease both males' and females' academic performance.

While the goal of the above research was not to see whether sex composition impinged upon educational performance and outcomes, some studies have probed into such a link. Unfortunately, the research to this point has not enabled scholars to reach a consensus. For example, Powell and Steelman (1989, 1990) found a "liability of having brothers"—in both high school grades (but not standardized test scores) and parental financial sponsorship for those in college. After weighing the merits of arguments gleaned from extensions of the confluence theory and resource dilution theories, Powell and Steelman attributed the first finding to another explanation—that the normative climate of the household is affected by sex composition. To the extent that girls earn higher grades in school than do boys, then youths with mostly sisters will be in an environment in which academic excellence is expected. This argument is, in some ways, a mirror image of the predictions emanating from Koch (1955) and Brim (1958).

The second finding regarding financial support of youths can be rooted in a variant of the resource dilution explanation. Contemporary evidence suggests that in subtle or not so subtle ways, parents, especially fathers, still express a preference for sons (Harris and Morgan 1991; Morgan, Lye, and Condran 1988). Moreover, financial returns on education may be greater for sons than for daughters, given the gender gap in lifetime earnings (Becker 1981). If favoritism for sons exists and/or if parents make "rational" calculations of returns on their own investments, parents may allocate more economic or other resources to their sons. If so, brothers are more potent claimants on resources than sisters. Brothers may prove to be more of an obstacle in other cultures, as in parts of East Asia where sisters have sacrificed their educational career and instead have entered the work force to help subsidize their brothers' education (Greenhalgh 1985).

The evidence regarding the harmful influence of brothers is inconclusive. In fact, Butcher and Case (1994) argued the opposite—that sisters provide the real obstacle to educational progress[1]—while Hauser and Kuo (1995) denied the presence of any sex composition effect on educational attainment.

Although it is difficult to reconcile these disparate findings, it is not impossible to merge some. For example, the findings from Powell and Steelman (1989, 1990) are not necessarily inconsistent with those from Hauser and Kuo (1995), since the former discussed educational performance in school (i.e., grades) and how college is funded, whereas the latter focused on educational attainment. Conceivably, sex composition has counterbalancing influences, or although educational attainment is not directly affected by sex composition, the routes through which education is attained are. Nevertheless, it will take careful thought and creative research to comb through this menagerie of possibilities.

Birth Order

While birth order has no obvious counterpart in the small groups literature, the age differences between siblings may set up a power structure in which age or ordinal position acts as a status characteristic. The implications with respect to birth order and educational outcomes are less certain than those associated with family size, but less murky than those connected to sex composition.

For some scholars, the notion that birth order has profound effects on personality and education has genuine appeal. Although judgments regarding birth order's link to personality hang in suspension (Ernst and Angst 1983; Sulloway 1995), the verdict regarding intelligence should be reaching closure. Although there have been hints of a birth order effect in which later-born youths are intellectually disadvantaged (Belmont and Marolla 1973; Paulhus and Shaffer 1981), most contemporary researchers have found meager evidence to support the link between birth order and intelligence once relevant confounding factors such as socioeconomic status and family size are taken into account (Blake 1989; Ernst and Angst 1983; Hauser and Sewell 1985; Olneck and Bills 1979; Page and Grandon 1979; Retherford and Sewell 1991; Steelman 1985; Steelman and Mercy 1980; Steelman and Powell 1985).

In terms of resource allocation, the effects of birth order are a mixed bag—they seem to depend on type of resource evaluated. Many commonsensical notions about birth order assume that a disproportionate share of resources is diverted to first-born children. This may hold true for parental time, energy, and interest as measured by, for example, whether pre-school aged children were read to by their parents and whether adolescents frequently talked with their parents about school (Powell and Steelman 1990, 1993). While one might surmise that earlier-born children are the beneficiaries of more economic resources, the research conducted in the United States generally finds otherwise. Although parents typically maintain higher aspirations for earlier-born children, later-born children are more likely to attend private school, have educational materials in the home, and receive financial assistance from their parents if they choose to attend post-secondary school (Powell and Steelman 1993; Steelman and Powell 1991). The upshot of this differential allocation of resources may be countervailing, thus yielding no overall impact of birth order on intellectual performance or educational attainment.

Sibling Spacing

Often overlooked, the relationship between the closeness in ages among siblings and children's intellectual development also flows directly from the theoretical views highlighted in the foregoing discussion on sibship size. Nonetheless, the paucity of information about birth intervals has probably contributed to our primitive knowledge about its impact. The sparse research that has been conducted suggests a deleterious impact of spacing on educational performance and attainment (Dandes and Dow 1969; Gailbraith 1982; Powell and Steelman 1990, 1993). Spacing between children also provides an instructive test for those interested in the research dilution hypothesis. Holding other factors constant, such as size and socioeconomic status, we might still expect educational attainment to decline as the spaced intervals between children shrink because allocation of parental resources may be constrained under such circumstances. For example, parental financial outlays may be made more difficult for pursuits, such as going to college or private schools, if children are closely spaced (Powell and Steelman 1993, 1995). Less clear is whether the same reasoning would extend to more social inputs. While there is no apparent counterpart for this variable in the small groups literature, we include it to complete our discussion of sibling configuration.

FUTURE DIRECTIONS IN FAMILIAL STRUCTURE RESEARCH

Predicting, or even recommending, future directions for sibship structure research is an arduous task. We already have identified several areas which need further elaboration. Surely there is a need to contemplate more seriously the meaning of resources and evaluate the mechanisms and forms under which the allocation of resources is determined by sibship size, sex composition, ordinal position, and spacing. Heightened attention to the distinctions among and the relative influences of social capital, economic capital, and culture capital also may prove instructive. More research is needed to determine whether there are critical junctures in the life cycles of children (and parents) in which resources play particularly salient roles. Further, although strong effects of some components of sibship structure, namely sex composition and birth order, on educational attainment has not been consistently documented, we still should consider the possibility that sibship features alter the route toward the same attainment. Another intriguing concept has been posited—negative resource dilution (Downey 1995). This represents the flip side of the resource dilution and challenges the assumption that parental resources are uniformly positive. Rather, this concept is based on the question: does the addition of siblings in the household diminish the likelihood of receiving negative resources—that is, resources which may penalize the child? For example, if a parent is abusive, does the presence of other siblings help shield the child from abuse? Or if a parent in the household does not speak English, does the presence of other siblings facilitate verbal skills? This real possibility is ignored in a literature that tends to extol the family unit as uniformly virtuous.

Fortunately, we now have data, such as High School and Beyond and the National Educational Longitudinal Study of 1988, which can help us reach greater resolution of the above issues. Unfortunately, even rich data sources such as these are incomplete in their attention to sibship structural variables. High School and

Beyond lacks information on sex composition while the National Educational Longitudinal Study includes no questions regarding sibship spacing.

It perhaps has become a cliché in review pieces such as this to encourage cross-cultural research. In this case, however, the rationale behind the cliché is defensible. Social psychologists have increasingly recognized the utility of testing theory across various cultural settings (Foschi 1980; Miller-Loessi 1995), although some social psychologists remain resistant to this viewpoint. The group processes literature has lagged behind other social psychological work in moving beyond lab experiments conducted in the United States, although Willer and Szmatka (1993), among others, offer an appreciated departure from this practice.

The family structure research probably has been more attuned to expanding its scope beyond the United States than has the small groups research. In fact, the confluence model was originally derived using Belmont and Marolla's (1973) data on test performance of young Dutch males. Research from other countries has been particularly insightful in identifying the micro and macro conditions under which sibship structure's effects may be magnified or lessened. Shavit and Pierce (1991), for example, found that the relationship between sibship size and educational attainment was not as clear-cut in Israel as it was in the United States. Although educational attainment decreased as a function of sibship size for Ashkenazi and Oriental Jews in Israel, the same pattern did not manifest itself for Moslem Arabs. Rather, for Moslem Arabs, the size of the hamula (i.e., the extended kin network) was critical. Shreeniwas' (1994) study of educational attainment in Malaysia also contested the universality of the sibship structure effects found in the United States. She documented that ethnic groups were differentially affected by sibship structure, which she attributed to the governmental policies which ensured higher educational participation rates for some ethnic groups. Others also have shown how governmental policies and cultural-specific views regarding gender, primogeniture, and the family can nullify or exacerbate the implications of family structure (Greenhalgh 1985; Parish and Willis 1993; Pong and Post 1995). Interestingly, while the effects of sibship size in the United States are unusually robust, the strongest effects of sex composition and, to a lesser degree, birth order have been found elsewhere in the international arena.

Although it is difficult to test whether the principles of sibship structure apply to other historical periods, some efforts have offered important insight. Clearly, we should not expect that the patterns regarding sibship structure should be constant across historical periods. Caldwell (1982), for example, convincingly argued that the negative effects of large family sizes should appear only after demographic transition has occurred. Moreover, perspectives regarding primogeniture and the relative value of daughters and sons may have changed. As an illustration, Butcher and Case (1994) credited the over-time decline in the effect of sex composition of sibship on girls to changing parental expectations and aspirations for their daughters.

Thus, the challenge for historically driven research on family structure is to identify the circumstances under which sibship structure is and is not important and to determine whether some effects of sibship structure have remained stable. Sulloway's (in preparation) historical account of the role of birth order on scientific risk taking is an exciting example of this genre of research. Using Galton's (1874) view on the scientific eminence of first-borns as a starting point, Sulloway turns this argument on its head and contends that later-born children are actually more receptive to novel ideas that revolutionize scientific thinking, such as Darwinian evolution, Copernican cosmology, and Hutton's theory of the Earth. While this is one of the boldest pronouncements regarding the influence of birth order on personality and intellectual processes, we must defer judgment on Sulloway's claims until his work is available for closer scrutiny.

THE WALL BETWEEN SMALL GROUPS AND FAMILY RESEARCHERS

Why has there been a lack of communication between those identified as small groups scholars and those identified as family ones? There seems to be an invisible wall between these two camps. In terms of historical development, oddly enough, the small groups research area began by concentrating on many of the same variables, such as the size of the group, that now engage those interested in family configuration. Simmel's (1950) classic work on groups, for example, showcased size as a pre-eminent variable impinging on interaction. Yet, these two streams with similar cross-currents usually fail to cross paths.

Surely both areas have great strengths, but with these regrettably come problems that have precluded potentially profitable dialogue between them. Small groups researchers should be praised for making tremendous theoretical advancements in their own realm—a feat that may have been possible only with focused and programmatic effort. With this narrowing scope, however, the possibility of screening out other orientations naturally arises. Small group researchers for the most part have relied upon data collected on unacquainted persons in laboratory settings (Heise 1995). This methodological preference was chosen in part to eliminate the contamination of noise (e.g., shared history and emotional connections). Testing theories without the clutter of this and other confounding variables lends itself most readily to experimental methods. This methodological decision was wise because it generated an enviable roster of accomplishments. Nevertheless, the small groups research, with its elegant theories and rigorous methodological orientation, may have become insulated from research outside its own sphere.

By contrast, family research is not as circumscribed as that on small groups. It is scattered across numerous locations in sociology (e.g., demography, sociology of education, and social stratification) and goes beyond disciplinary confines (e.g., sociology, evolutionary biology, economics, and psychology). In addition, family research has not been particularly successful bridging the gap between applied and theoretical analyses, and, as a result, it has been less consistently driven by theory than has small groups research. A core theoretical thrust has, thus, not evolved.

Moreover, the methodological approaches used in family research, while not diametrically opposed to those employed by small group researchers, certainly are more diverse—including, but not limited to, surveys, observational studies, and case studies. The family is a group ill-suited to examination in the laboratory. The family carries with it much noise—it has a history, familiarity, and emotional intimacy. For the family researcher, thus, the task of eliminating contamination from inspection is complicated; however, many would philosophize that it is precisely the noise of the family which should beckon the most attention. Nonetheless, while the literature on the family may be rightly criticized for being unfocused, in part because it is so theoretically and methodologically diffuse, at the very least it has attempted to address the group which may link the individual to society in the most profound way. Perhaps because of its visibility across an array of fields, the somewhat eclectic approach taken in studying it, the failure of those who study it to develop a core set of ideas, and the wide audience it attracts, the study of the family is not an easy one to fuse with work on small groups.

Certainly we are not implying that small groups theory has completely ignored the family or vice versa. To the contrary, exchange theory, for example, has been used somewhat fruitfully in literature on dating, mate selection, and power imbalance between spouses (Molm and Cook 1995); however, the theory's potential contributions to understanding the implications for most areas of family sociology—including the study of sibship structure and resource allocation—have not been explored.

Still there are some fine examples that integrate small group theory and family research. In a provocative paper, Bonacich, Grusky, and Peyrot (1985) drew upon the coalition theories formulated by Caplow (1968) and Gamson (1961), two early contributors to small groups research, and applied them to the family. In this piece, they found that coalitions operated to maintain the status of the more powerful members of the family—that is, parents. In another rare application of small group theory tested in the family setting, Felson and Russo (1988) found that parents more likely punished their children deemed as more powerful, defined as older and male siblings—results they interpreted as somewhat contradictory to previously mentioned studies on coalitions. They also found that intervening on behalf of the weaker party actually results in greater aggression. These illustrations show that some social scientists have flirted with consolidating ideas across these branches of research; however, such creativity seems to have stalled or, at best, is rarely displayed.

As demonstrated from the research of sibship structure, more interplay between the two camps is not only feasible but desirable. The basic idea that groups possess a character beyond the individuals who comprise it is a central tenet undergirding much social psychological thought. Since families are structured in different ways, there may well be systematic differences for their members that transcend any of their idiosyncratic traits. In parallel fashion, the behavior of lab subjects may be altered by the structural features of the groups to which they are exposed regardless of the individual differences they bring into the laboratory situation. If similar patterns are observed across the continuum of intimacy from the family to the small group, then we could move more surefootedly toward general principles of group behavior.

Our position actually squares with the stance outlined by some scholars who study small groups and group processes. Willer, Lovaglia, and Markovsky (1995), for example, underscored the value of applying network exchange theory beyond the small group in order to understand more broadly macro arenas. And nearly every chapter on small group processes in the recent *Sociological Perspectives on Social Psychology* (Cook, Fine, and House 1995) at least made the obligatory recommendation that lessons learned from small groups research be extended outside the experimental setting.

Whether there is convergence between theory and the research that is presently being conducted on the family and on small groups remains to be seen. Regarding the link between sibship structure and resource allocation—a topic we highlighted in this paper—one can immediately think of several questions emanating from the small groups and group processes literature: Do diffuse status characteristics in the family (such as sex, age, and ordinal position) affect perceptions of competence and, in turn, resource allocation in the family? When parents make decisions regarding the provision of resources, which is paramount—equity, equality, or need? How do size of family, sex composition, age spacing, and ordinal position influence perceptions of distributive justice in the family? How does sibship structure affect the types of and amount of networks youths have outside of the family? Do the networks that youths possess outside of the family (and inside the family) shape the distribution of resources within the family?

At some point, the curious scientist will want to understand the conditions under which theoretical principles of group processes apply to a real small group and the conditions under which they do not. As we have expressed beforehand, if common principles can be found that characterize groups from the family to the lab group, then our confidence in such theorizing would be greatly enhanced. However, if common ground is not uncovered, we must ask: what distinguishes the familial groups from other groups so that the general principles of groups processes do not apply to the family? Thus, even if no convergence is unearthed, it then becomes incumbent upon social scientists to try to figure out why.

ACKNOWLEDGMENTS

An earlier version of this paper was presented at the annual meeting of the American Sociological Association in Pittsburgh, August 1992. The authors contributed equally to this paper. This project was supported in part by the Institute for Families in Society, University of South Carolina, and the Office of Academic Affairs and Dean of Faculties, Indiana University. We would like to thank Robert Carini, Douglas Downey, Carl Ek, Christina Ek, Robert Fulk, Jennifer Stewart, Sheldon Stryker, and David Willer for their helpful comments and support.

NOTE

1. This discussion does not consider all of the different possibilities regarding the effects of sex composition on educational achievement and attainment. Butcher and Case (1994) listed several others that should be mentioned. Using a reference group theory, one might contend that in households in which youths have no same-sex sibling, they will use their opposite-sex sibling as their referent, as their rival. However, the arrival of a same-sex sibling would alter this reference and, consequently, affect educational performance. Using some strands of equity theory, one might predict that parents will invest more in their daughters precisely because of sex inequality in wages. Alternatively, sex composition may influence parental investments because the costs, both financially and emotionally, of raising sons vs. daughters may vary. If the total cost of raising a daughter, for example, is more expensive, then youths (both males and females) with several sisters will be in a relatively disadvantageous position. Another possibility, not posed by Butcher and Case, is that parents will be in a more financially advantageous position if they have children of one sex only. Having sons only or daughters only may be less expensive for parents because they will have to spend less money on certain items (for example, clothing) which can be passed on to each child. If so, then being in a single-sex sibship is beneficial for children because there will be more discretionary funds available.

REFERENCES

Adler, A. 1959. *The Individual Psychology of Alfred Adler: A Systematic Presentation in Selections From His Writings*, edited and annotated by H. L. Ansbacher and R. R. Ansbacher. New York: Basic Books.
Alwin, D. F. 1984. "Trends in Parental Socialization Values: Detroit, 1958-1983." *American Journal of Sociology* 90: 359-382.
Alwin, D. F., and A. Thornton. 1984. "Family Origins and the Schooling Process: Early Versus Late Influence of Parental Characteristics." *American Sociological Review* 49: 784-802.
Amato, P., L. S. Loomis, and A. Booth. 1995. "Parental Divorce, Marital Conflict, and Offspring Well-Being During Early Adulthood." *Social Forces* 73: 895-915.
Anastasi, A. 1956. "Intelligence and Family Size." *Psychological Bulletin* 53: 187-209.
Bales, R. F., A. P. Hare, and E. F. Borgatta. 1957. "Structure and Dynamics of Small Groups: A Review of Four Variables." Pp. 391-422 in *Review of Sociology: Analysis of a Decade*, edited by J. B. Gittler. New York: Wiley.
Becker, G. S. 1964. *Human Capital*. Cambridge, MA: Harvard University Press.
_____. 1981. *A Treatise on the Family*. Cambridge, MA: Harvard University Press.
Becker, G. S., and N. Tomes. 1976. "Child Endowments and the Quantity and Quality of Children." *Journal of Political Economy* 84: S143-S162.
Belmont, L., and F. A. Marolla. 1973. "Birth Order, Family Size and Intelligence." *Science* 182: 1096-1101.
Blake, J. 1981. "Family Size and the Quality of Children." *Demography* 18: 421-442.
_____. 1986. "Number of Siblings, Family Background, and the Process of Educational Attainment." *Social Biology* 33: 5-21.
_____. 1989. *Family Size and Achievement*. Berkeley, CA: University of California Press.
Blau, P., and O. D. Duncan. 1967. *The American Occupational Structure*. New York: Wiley and Sons.

Bonacich, P., O. Grusky, and M. Peyrot. 1985. "Family Coalitions: A New Approach and Method." *Social Psychology Quarterly* 48: 42-50.

Bossard, J. H. S., and E. S. Boll. 1956. *The Large Family System*. Philadelphia, PA: University of Pennsylvania Press.

Bourdieu, P. 1977. *Reproduction in Education, Society and Culture*. Beverly Hills, CA: Sage.

Brim, O. G. 1958. "Family Structure and Sex Role Learning by Children." *Sociometry* 21: 1-16.

Brody, C. J., and L. C. Steelman. 1985. "Sibling Structure and Parental Sex-Typing of Children's Household Tasks." *Journal of Marriage and the Family* 45: 265-273.

Broman, S. H., Paul L. Nichols, and W. A. Kennedy. 1975. *Preschool IQ: Prenatal and Early Developmental Correlates*. London: Earlbaum.

Bumpass, L. L. 1984. "Children and Marital Disruption: A Replication and Update." *Demography* 21: 71-82.

Bumpass, L. L., R. R. Rindfuss, and R.B. Janosik. 1978. "Age and Marital Status at First Birth and the Pace of Subsequent Fertility." *Demography* 15: 75-86.

Butcher, K., and A. Case. 1994. "The Effect of Sibling Sex Composition on Women's Education and Earnings." *Quarterly Journal of Economics* 109: 531-563.

Caldwell, J. C. 1982. *Theory of Fertility of Decline*. New York: Academic Press.

Caplow, T. 1968. *Two Against One: Coalitions in Triads*. Englewood Cliffs, NJ: Prentice-Hall.

Cicirelli, V. G. 1978. "The Relationship of Sibling Structure to Intellectual Abilities and Achievements." *Review of Educational Research* 55: 353-386.

Coleman, J. S. 1988. "Social Capital in the Creation of Human Capital." *American Journal of Sociology* 94: S95-S120.

Cook, K. S., G. A. Fine, and J. S. House. Eds. 1995. *Sociological Perspectives on Social Psychology*. Needham Heights, MA: Allyn and Bacon.

Cooley, C. H. 1909. *Social Organization*. New York: The Free Press.

Dandes, H. M., and D. Dow. 1969. "The Relation of Intelligence to Family Size and Density." *Child Development* 40: 641-644.

DiMaggio, P. 1982. "Cultural Capital and School Success: The Impact of Status Culture Participation on the Grades of U.S. High School Students." *American Sociological Review* 47: 189-210.

Downey, D. B. 1995. "Bigger Is Not Better: Family Size, Parental Resources, and Children's Educational Performance." *American Sociological Review*.

Downey, D. B., P. B. Jackson, and B. Powell. 1994. "Sons Versus Daughters: Sex Composition of Children and Maternal Views on Socialization." *Sociological Quarterly* 35: 33-50.

Downey, D. B., and B. Powell. 1993. "Do Children in Single-Parent Households Fare Better Living With Same-Sex Parents?" *Journal of Marriage and the Family* 55: 55-71.

Elder, G. H. 1985. *Life Course Dynamics: Trajectories and Transitions, 1968-1980*. Ithaca, NY: Cornell University Press.

Elder, G. H., and C. E. Bowerman. 1963. "Family Structure and Childrearing Patterns: The Effect of Family Size and Sex Composition." *American Sociological Review* 28: 891-905.

Ernst, C., and J. Angst. 1983. *Birth Order: Its Influence on Personality*. Berlin, Germany: Springer-Verlag.

Featherman, D., and R. Hauser. 1978. *Opportunity and Change*. New York: Academic Press.

Felson, R. B., and N. Russo. 1988. "Parental Punishment and Sibling Aggression." *Social Psychology Quarterly* 51: 11-18.

Foschi, M. 1980. "Theory, Experimentation and Cross-Cultural Comparisons in Social Psychology." *Canadian Journal of Sociology* 5: 91-102.

Foschi, M., and E. J. Lawler. 1994. *Group Processes: Sociological Analyses*. New York: Nelson-Hall.

Furstenberg, F. F., and A. J. Cherlin. 1991. *Divided Families: What Happens to Children When Parents Part?* Cambridge, MA: Harvard University Press.

Gailbraith, R. C. 1982. "Sibling Spacing and Intellectual Development: A Closer Look at the Confluence Model." *Developmental Psychology* 18: 151-174.

Galton, F. 1874. *English Men of Science: Their Nature and Nurture.* London: Macmillan.

Gamson, William. 1961. "A Theory of Coalition Formation." *American Sociological Review* 26: 373-382.

Goldscheider, F. K., and C. Goldscheider. 1989. *Ethnicity and the New Family Economy: Living Arrangements and Intergenerational Financial Flows.* Boulder, CO: Westview.

Gomes, M. 1984. "Family Size and Educational Attainment in Kenya." *Population and Development Review* 4: 647-660.

Greenhalgh, S. 1985. "Sexual Stratification: The Other Side of 'Growth Versus Equity' in East Asia." *Population and Human Development Review* 11: 265-314.

Grotevant, H. D., S. Scarr, and R. Weinberg. 1977. "Intelligence and Development in Family Constellations With Adopted and Natural Children: A Test of the Zajonc and Markus Model." *Child Development* 48: 1699-1703.

Grusky, O., P. Bonacich, and M. Peyrot. 1988. "Group Structure and Interpersonal Conflict in the Family." Pp. 29-31 in *Advances in Group Processes,* Vol. 5, edited by E. Lawler and B. Markovsky. Greenwich, CT: JAI Press Inc.

Guttentag, M., and P. F. Secord. 1983. *Too Many Women: The Sex Ratio Question.* New York: Sage Publications.

Hare, A. P. 1952. "A Study of Interaction and Consensus in Different Sized Groups." *American Sociological Review* 17: 261-267.

Harris, K. M., and S. P. Morgan. 1991. "Fathers, Sons, and Daughters: Differential Paternal Involvement in Parenting." *Journal of Marriage and the Family* 53: 531-544.

Hauser, R. M., and H. H. Daphne Kuo. 1995. "Does the Gender Composition of Sibships Affect Educational Attainment?" Paper presented at the 1995 Meetings of the Population Association of America, San Francisco, CA.

Hauser, R. M., and W. H. Sewell. 1985. "Birth Order and Educational Attainment in Full Sibships." *American Educational Research Journal* 22: 1-23.

Hdry, S. B. 1987. "Sex-Biased Parental Investment Among Primates and Other Mammals: A Critical Evaluation of the Trivers-Willard Hypothesis." Pp. 97-148 in *Child Abuse and Neglect: Biosocial Dimensions,* edited by R. J. Gelles and J. B. Lancaster. New York: Aldine De Gruyter.

Heer, D. M. 1985. "Effects of Sibling Number on Child Outcomes." *Annual Review of Sociology* 11: 27-47.

_____. 1986. "Effect of Number, Order, and Spacing of Siblings on Child and Adult Outcomes." *Social Biology* 33: 1-4.

Heise, D. 1995. "Review of Group Processes: Sociological Analyses." *Social Forces* 73: 1120-1121.

Heise, D., and E. P. M. Roberts. 1970. "The Development of Role Knowledge." Genetic Psychology Monographs 82: 83-115.

Hetherington, E. M. 1989. "Coping With Family Transitions: Winners, Losers, and Survivors." *Child Development* 60: 1-14.

Higgins, J.V., E. W. Reed, and S.C. Reed. 1963. "Intelligence and Family Size: A Paradox Resolved." *Eugenics Quarterly* 9: 84-90.

Hill, C. R., and F. P. Stafford. 1974. "Allocation of Time to Preschool Children and Educational Quality." *Journal of Human Resources* 9: 323-341.

Kanter, R. M. 1977. *Men and Women of the Corporation.* New York: Basic Books.

Koch, H. 1955. "Some Personality Correlates of Sex, Sibling Position, and Sex of Sibling Among Five and Six Year Children." *Genetic Psychology Monographs* 52: 3-50.

Kohn, M. L. 1969. *Class and Conformity: A Study in Values, Second Edition.* Chicago: University of Chicago Press.

Kohn, M. L., K. M. Slomczynski, and C. Schoenbach. 1986. "Social Stratification and the Transmission of Values in the Family: A Cross-National Assessment." *Sociological Forum* 1: 73-102.

Lindert, P. H. 1974. "Family Inputs and Inequality Among Children." Discussion paper No. 128-34. Madison: Institute for Research for Poverty, University of Wisconsin.

Mabry, E. A. 1985. "The Effects of Gender Composition and Task Structure on Small Group Interaction." *Small Group Behavior* 16: 75-97.

Marjoribanks, K. 1972. "Environment, Social Class, and Mental Abilities." *Journal of Educational Psychology* 63: 103-109.

McLanahan, S., and K. Booth. 1989. "Mother-Only Families: Problems Prospects, and Politics." *Journal of Marriage and the Family* 30: 105-116.

Mercy, J. A., and L. C. Steelman. 1982. "Familial Influence on the Intellectual Attainment of Children. *American Sociological Review* 47: 532-543.

Miller-Loessi, K. 1995. "Comparative Social Psychology: Cross-Cultural and Cross-National." Pp. 396-420 in *Sociological Perspectives on Social Psychology*, edited by K. S. Cook, G. A. Fine, and J. S. House. Needham Heights, MA: Allyn and Bacon.

Molm, L. D., and K. S. Cook. 1995. "Social Exchange and Exchange Networks." Pp. 209-235 in *Sociological Perspectives on Social Psychology*, edited by K. S. Cook, G. A. Fine, and J. S. House. Needham Heights, MA: Allyn and Bacon.

Morgan, S. P., D. N. Lye, and G. Condran. 1988. "Sons, Daughters, and Divorce: Does the Sex of the Child Affect the Risk of Marital Disruption?" *American Journal of Sociology* 94: 110-130.

Nisbet, J.D., and N.J. Entwistle. 1967. "Intelligence and Family Size: 1949-1965." *British Journal of Educational Psychology* 37: 188-193.

Nye, I. F., J. Carlson, and G. Garrett. 1970. "Family Size, Interaction, Affect, and Stress." *Journal of Marriage and the Family* 32: 216-226.

Olneck, M. R., and D. B. Bills. 1979. "Family Configuration and Achievement: Effects of Birth Order and Family Size in a Sample of Brothers." *Social Psychology Quarterly* 42: 135-148.

Page, E. B., and G. M. Grandon. 1979. "Family Configuration and Mental Ability: Two Theories Contrasted With U.S. Data." *American Educational Research Journal* 16: 257-272.

Parish, W. L., and R. J. Willis. 1993. "Daughters, Education, and Family Budgets: Taiwan Experiences." *Journal of Human Resources* 28: 863-898.

Paulhus, D., and D. R. Shaffer. 1981. "Sex Differences in the Impact of Number of Older and Number of Younger Siblings on Scholastic Aptitude." *Social Psychology Quarterly* 44: 363-368.

Pong, S. L., and D. Post. 1995. "Children's Educational Attainment in Lima and Hong Kong: Changing Impact of Sex Composition, Sib Size and Birth Order." Paper presented at the 1995 Meetings of American Sociological Association, Washington, DC.

Powell, B., and L. C. Steelman. 1989. "The Liability of Having Brothers: Paying for College and the Sex Composition of the Family." *Sociology of Education* 62: 134-147.

_____. 1990. "Beyond Sibship Size: Sibling Density, Sex Composition, and Educational Outcomes." *Social Forces* 69: 181-206.

_____. 1993. "The Educational Benefits of Being Spaced Out: Sibship Density and Educational Progress." *American Sociological Review* 58: 367-381.

_____. 1995. "Feeling the Pinch: Child-Spacing and Constraints on Parental Economic Investments in Children." *Social Forces*.

Rankin, R. J., A. J. Gaite, and T. Heiry. 1979. "Cultural Modification of the Effects of Family Size on Intelligence." *Psychological Reports* 45: 391-397

Retherford, R. D., and W. H. Sewell. 1991. "Birth Order and Further Tests of the Confluence Model." *American Sociological Review* 56: 141-158.

Ridgeway, C. L., 1988. "Gender Differences in Task Groups: A Status and Legitimacy Account." Pp. 188-206 in *Status Generalization*, edited by M. A. Webster, Jr., and M. Foschi. Stanford, CA: Stanford University Press.

Rosen, B. C., 1961. "Family Structure and Achievement Motivation." *American Sociological Review* 26: 574-584.

Sewell, W. H., R. M. Hauser, and W. Wolf. 1980. "Sex, Schooling and Occupational Status." *American Journal of Sociology* 86: 551-583.

Shavit, Y., and J. L. Pierce. 1991. "Sibship Size and Educational Attainment in Nuclear and Extended Families." *American Sociological Review* 56: 321-330.

Shreeniwas, S. 1994. *Status Attainment in Peninsular Malaysia: The Impact of Familial Socio-Demographic Characteristics and State Politics Over Times.* Unpublished dissertation. University of Michigan.

Simmel, G. 1950. *The Sociology of Georg Simmel,* translated by Kurt H. Wolff. Glencoe, IL: Free Press.

Slater, P. E. 1958. "Contrasting Correlates of Group Size." *Sociometry* 21: 129-139.

Smith-Lovin, L., J. V. Skvoretz, and C. G. Hudson. 1986. "Status and Participation in Six-Person Groups: A Test of Skvoretz's Comparative Model." *Social Forces* 64: 992-1005.

Steelman, L. C. 1985. "A Tale of Two Variables: The Intellectual Consequences of Sibship Size and Birth Order." *Review of Educational Research* 55: 353-386.

Steelman, L. C., and J. A. Mercy. 1980. "Unconfounding the Confluence Model: A Test of Sibship Size and Birth Order Effects on Intelligence." *American Sociological Review* 45: 571-582.

Steelman, L. C., and B. Powell. 1985. "The Social and Academic Consequences of Birth Order: Real, Artifactual, or Both?" *Journal of Marriage and the Family* 47: 117-124.

_____. 1989. "Acquiring Capital for College: The Constraints of Family Configuration." *American Sociological Review* 54: 844-855.

_____. 1991. "Sponsoring the Next Generation: Parental Willingness to Pay for Higher Education." *American Journal of Sociology* 96: 1505-1529.

Sulloway, F. J. 1995. "Birth Order and Evolutionary Psychology: A Meta-Analytic Overview." *Psychological Inquiry: An International Journal* 6: 75-80.

_____. In preparation. *Born to Rebel: Radical Thinking in Science and Social Thought.*

Taubman, P., and J. R. Behrman, 1986. "Effect of Number and Position of Siblings and Adult Outcomes." *Social Biology* 33: 22-33.

Teachman, J. R. 1987. "Family Background, Educational Resources, and Educational Attainment." *American Sociological Review* 52: 548-577.

Teachman, J. R., and P. T. Schollaert. 1989. "Gender of Children and Birth Timing." *Demography* 26: 411-422.

Wiley, M. G. 1995. "Sex Category and Gender in Social Psychology." Pp. 362-386 in *Sociological Perspectives on Social Psychology,* edited by K. S. Cook, G. A. Fine, and J. S. House. Needham Heights, MA: Allyn and Bacon.

Willer, D., M. Lovaglia, and B. Markovsky. 1995. "Power and Influence." Paper presented at the 1995 Meetings of the American Sociological Association, Washington, DC.

Willer, D., and J. Szmatka. 1993. "Cross-National Experimental Investigations of Elementary Theory: Implications for the Generality of the Theory and the Autonomy of Social Structure." Pp. 37-82 in *Advances in Group Processes,* Vol. 10, edited by E. Lawler, B. Markovsky, K. Heimer, and J. O'Brien. Greenwich, CT: JAI Press Inc.

Zajonc, R. B., and G. Markus. 1975. "Birth Order and Intellectual Development." *Psychological Review* 82: 74-88.

METAMETHODOLOGY AND PROCEDURAL DECISIONS IN THEORETICAL EXPERIMENTATION:
THE CASE OF POWER PROCESSES IN BARGAINING

Rebecca Ford

ABSTRACT

This paper provides a metamethodological analysis of the experimental setting used to evaluate bilateral deterrence and conflict spiral theories, two theories of punitive power in explicit bargaining. A core concern is how explicated metatheory and theory shape procedural decisions such as type of experiment, strategies for operationalizing independent variables, measuring dependent variables, controlling non-relevant variables, and instantiating scope conditions for the theories. This implies that metatheoretical ideas about how power, conflict, bargaining, and the actor should be conceptualized will impact procedural decisions as will specific independent variables, dependent variables, and scope conditions specified by the theory. Analysis of linkages between concrete, empirical features of a laboratory setting and the more abstract metatheory and theory that it evaluates is important because procedural decisions remain under analyzed, often appearing as intuitive and ad hoc decisions, and because laboratory settings tend to represent implicit or explicit statements of conditionalization for programs of research. Both reasons present problems for interpreting findings and assessing cumulation over a series of studies.

Advances in Group Processes, Volume 13, pages 237-263.
Copyright © 1996 by JAI Press Inc.
All rights of reproduction in any form reserved.
ISBN: 0-7623-0005-1

INTRODUCTION

This paper provides a metamethodological analysis of the experimental paradigm used in a theoretical program of research on power processes in bargaining (see Bacharach and Lawler 1981; Lawler 1986; Ford and Blegen 1992; Lawler and Bacharach 1986, 1987; Lawler, Ford, and Blegen 1988). The primary theoretical question addressed by this program of research is how the power capabilities of two parties in conflict influence their use of that power during explicit bargaining. Explicit bargaining entails acknowledgment of a bargaining relationship, open exchange of provisional offers, and an issue that permits intermediate solutions (Chertkoff and Esser 1976). Power is conceptualized as a nonzero sum, structurally based capability; power use involves tactical action, hostile or conciliatory, that can affect outcomes of another. Conciliatory tactics take the form of concessions and agreements, while hostile tactics take the form of threats and damage (Bacharach and Lawler 1981; Lawler 1992; Lawler and Ford 1993).

The program consists of two distinct branches of theory and associated research that focus on dependence power and punitive power respectively (e.g., Bacharach and Lawler 1981, chapter 3; Boyle and Lawler 1991; Emerson 1962; 1972; Ford and Blegen 1992; Lawler and Bacharach 1987; Lawler, Ford, and Blegen 1988). While analyses of the two forms of power have been parallel and complementary, growth in the punitive power branch has been somewhat more complex and interesting. For this reason, I use it to illustrate how standardized experimental procedures developed to test theories of punitive power are shaped by both theoretical and metatheoretical concerns.[1]

Previous analysis of the power processes research program concentrated on identifying the program's metatheoretical core, its link to the two punitive power theories, bilateral deterrence and conflict spiral theory, and the core's importance for subsequent theoretical growth (Lawler 1992; Lawler and Ford 1993). The term *metatheory* refers to sets of epistemological and ontological assumptions that (1) provide a problem focus, and (2) establish guidelines for conceptualizing key theoretical constructs, for example, power, conflict, and bargaining. The former are termed *orienting assumptions*, while the latter are the *metatheoretical core* (Lawler and Ford 1993)." In contrast, the term *theory* applies to abstract, conditionalized claims that are testable in principle (Wagner 1984). Bilateral deterrence and conflict spiral theories are both social psychological theories that predict the use of punitive power in bargaining but offer contradictory predictions about power magnitude and degree of power equality, for example, bilateral deterrence predicts that larger levels of equal power lead to less use of punitive tactics, while conflict spiral theory predicts the opposite.

We also analyzed the importance of *friendly competition* between bilateral deterrence and conflict spiral theories to stimulating parallel theoretical developments. Friendly competitors are theoretical variants, for example, theories that contain both common and contradictory elements (see Wagner 1984; Wagner and

Berger 1985). This type of relationship between theories should push parallel theoretical elaboration (i.e., increases in scope, rigor, precision, or empirical adequacy) rather than seeking to eliminate one or the other (Wagner 1984).[2]

In this paper, I continue analysis of the power processes research program by examining how metatheory, theory, and friendly competition shape theory evaluation decisions(see also Berger, Fisek, Norman, and Zelditch 1977). Theory evaluation and other procedural decisions in a program of work are influenced in crucial ways by metatheoretical assumptions about the nature and goals of scientific knowledge, desirable forms of theorizing, and the substantive problem focus. For example, metatheories that value historical, context-dependent knowledge and the non-reducibility of human thought and action are more likely to imply descriptive theorizing and an interpretative, narrative, or ethnographic method (see, e.g., Gergen 1973, 1978, 1985; Gubrium 1988; Holstein and Gubrium 1995). Metatheories emphasizing abstract, general, conditional, and formal knowledge and a simplified model of the actor are more likely to imply experimental or other quantitative methods (e.g., Cohen 1980; Lawler, Ridgeway, and Markovsky 1993). Thus, varying presuppositions about desirable and possible forms of knowledge produce varying kinds of data and tell us different things about social reality.

Because the power processes program relies on laboratory experimentation to evaluate theoretical predictions, I focus on the links between theory *and metatheory* in shaping procedural decisions in a laboratory setting. While most of my analysis focuses on the link between substantive metatheoretical and theoretical concepts and assumptions and experimental procedures, the analysis also notes some linkages between experimental procedures and the strategy of friendly competition.

A more abstract analysis of experimental procedures is important for two interrelated reasons. First, relative to design and statistical issues, procedural decisions remain under analyzed and have been described as a largely "artistic" or "intuitive" process lacking firm guidelines (Aronson, Brewer, and Carlsmith 1985; Aronson and Carlsmith 1968).[3] However, even the simplest settings involve many complex decisions about how to best translate an abstract idea into concrete features (e.g., kind of task, instructions, instantiation of independent variables, types of measures, etc.). Such decisions are often seemingly made in an ad hoc fashion, with no clear theoretical rationale, or no obvious connection to the ideas being tested. Given that abstract ideas can be tested in many different ways, failure to explicate the metatheoretical and theoretical reasoning behind such decisions increases the likelihood of multiple interpretations and decreases the likelihood of recognizing the inseparability of theory and data (Aronson, Brewer, and Carlsmith 1985; Aronson and Carlsmith 1968; Secord 1977; Willer 1987).[4,5]

Second, different research traditions have developed distinctive experimental settings tied to the specific test conditions required for a given theory or theoretical perspective. These settings are themselves social contexts that create varying definitions of the situation and so represent implicit scope conditions (i.e., the abstract conditions under which a theory is presumed to hold) for associated theory and research (Lawler and Ford 1994; Secord 1977; Stryker 1977, 1989; Walker and Cohen 1985). As applied to the bargaining area, this would suggest that matrix games that present dichotomous, simultaneous choices create different social contexts (i.e., tacit bargaining) than do bilateral monopoly negotiation games that allow open communication over a range of possible solutions (i.e., explicit bargaining) and create different definitions of the situation for subjects. Failure to recognize such differences as implicit statements of conditionalization for programs of work may lead to apparent inconsistencies in findings or unwarranted comparisons between different research traditions (Lawler and Ford 1995). Thus, a more abstract analysis of the features of experimental settings and their links to theory and metatheory has important implications for understanding both the generality and limitations of experimental findings and for assessing cumulation (qualitatively or quantitatively) both within and across programs of research.

The remainder of the paper is divided into four sections. First, I describe some of the basic metamethodological assumptions driving the program of research. These include assumptions about the purpose of theoretical research, the nature and form of theories of punitive power, the role of metatheory in theoretical growth, and the general import of these assumptions to use of laboratory settings (See Lawler 1992; Lawler and Ford 1993 for a more complete description of these concepts and how they shape theory construction).

This discussion is linked to a brief contrast between theoretical and empirical experimentation. While this distinction is not new (see, e.g., Cohen 1980, 1989; Freese 1980; Webster and Kervin 1971; Willer 1987), it is important to this paper for two reasons. First, the contrast underscores the distinctiveness of a theoretical approach to experimentation and suggests why metamethodological experimental strategies vary as they do. Second, the distinction suggests why empirical researchers may overlook implicit statements of scope associated with their experimental paradigms.

Second, I review the metatheoretical core and its link to procedural decisions. Much as these assumptions serve an orienting function in theory construction, they serve a similar role in theory evaluation; for example, they provide guidance about how to construct some of the most basic features of the experimental setting designed to test punitive power theories. Third, I review both punitive power theories (bilateral deterrence and conflict spiral), their most basic predictions, and several tests of the theories. Finally, I examine how punitive power theory and metatheory help design the experimental setting used to test bilateral deterrence and conflict spiral theories. This discussion is organized around the concrete kinds of control decisions that must be made in a punitive power experiment—for example, how to create and manipulate independent variables, how to create and measure dependent variables, how to control extraneous sources of variation, how to create and hold constant scope conditions for tests of the theories.

THEORETICAL VS. EMPIRICAL EXPERIMENTATION

Theoretical Experimentation

Strategies for evaluating theory are shaped in crucial ways by underlying metamethodological (i.e., epistemological) assumptions guiding a program of research (Berger, Wagner, and Zelditch 1992). Programs of research vary in the extent to which they articulate these assumptions and note their implications for the research process (see also Holstein and Gubrium 1995). Three core metamethodological assumptions in the power processes program include the following: (1) research should evaluate theoretical, that is, abstract, simplified, conditionalized accounts of the process by which punitive power capability is used during bargaining; (2) research should further revision and additional developments among theories of punitive power (see, e.g., Berger and Wagner 1985); (3) an explicit metatheory facilitates complex theoretical growth and guides theory assessment. Assumptions like these are typical of theoretical research programs that stress the importance of building and extending abstract conditionalized theories (see, e.g., Berger, Wagner, and Zelditch 1989, 1992; Berger and Zelditch 1993; Willer 1987).

Although other methods are certainly possible (depending on the dictates of the metatheory and theory), experimentation is often the method of choice among theoretical research programs with this vision of theory and metatheory because it provides the kind of control necessary to create and recreate the hypothetical situation dictated by conditionalized theory and close out processes not specified by theory. Thus, theoretical experimentation entails using control strategically to test, revise, and build theory.

Of course, empirical researchers find experimental control appealing as well. While empirical experimentation involves many of the same decisions and features as does theoretical experimentation—for example, developing the context, constructing the independent variable, measuring the dependent variable, controlling extraneous variables, ensuring and assessing impact of independent variables, and interpreting findings—it differs in purpose and in its strategy for constructing settings and interpreting findings (for a more complete analysis of theoretical experimentation, see Foschi 1980; Freese 1980; Henshel 1980; Webster and Kervin 1971; Willer 1987; Zelditch 1968).

Previous analyses of theoretical experimentation have argued that its purpose is to test basic theoretical implications, and to foster growth among abstract, scope constrained theories (Webster and Kervin 1971; Willer 1987). By implication, the laboratory setting is an artificial situation that is a function of the explicit, substantive metatheory guiding a program of research and the particular theory under test. Substantive metatheoretical assumptions, for example, those about the actor and processes of interest, provide guidance about such procedural decisions as type of experiment (e.g., impact vs. judgement; deception vs. nondeception), how to conceptualize and introduce independent variables and assess their impact, and which processes should be closed out of the setting. For example, cognitive metatheories that visualize the actor in subjective, interpretative terms and stress the importance of perceptual processes are more likely to rely on judgement experiments in which subjects describe their perceptions, attitudes, and motivations than behavioral metatheories which are more likely to favor impact experiments that produce behaviors and ignore internal states.

Other features of a setting are dictated by the actual requirements of the theory itself, for example, scope conditions, and specific independent and dependent variables (Cohen 1980; Foschi 1980). For example, bilateral deterrence and conflict spiral theories apply only to explicit bargaining contexts. This implies a laboratory setting in which actors recognize a difference of interest and can openly exchange a series of provisional offers along an issue continuum with multiple solutions. Thus, from this perspective, a prisoner's dilemma matrix[6] would not be an appropriate context in which to test theoretical implications of bilateral deterrence and conflict spiral theories because it does not clearly fit this definition of bargaining—despite the fact that it arguably represents a mixed motive situation.[7]

Theoretical experiments tend to be highly standardized and make use of the same cover story,[8] the same or similar operationalizations of independent variables, measure dependent variables similarly, and control theoretically irrelevant variables in the same way over a series of studies because this facilitates comparing one set of findings with another and provides a cleaner test of the theory in question (Berger, Fisek, Norman, and Zelditch 1977). Changes in the setting flow from basic theoretical implications or from modifications to a theory (e.g., elaboration, in which a second version of a theory is more precise or rigorous than the first [Wagner 1984]). For instance, some implications of bargaining theories suggest open interaction, bilateral monopoly contexts where actors actually exchange offers and counteroffers; others may suggest the need for programmed interaction where actors face standardized concession sequences.

Despite efforts to make as many features of a setting correspond to theoretical concepts as possible, there are always some variables present that do not link to the theory but which may impact the dependent variable—for example, age, sex, educational attainment, region of origin—and so must be controlled. These variables are not varied in a theoretical experiment unless they can be conceptualized in theoretical terms. An explicated metatheory may suggest which of these experimental limitations has theoretical potential and may be seen as an indicator of a theoretical concept. If this happens, both theory and setting are modified; if not, the importance of these variables tends to be minimized (Foschi 1980).

Finally, because theoretical experimenters do not link the laboratory setting to situations outside the laboratory, or regard the setting as an analog of anything outside the laboratory, findings are applied only to the evaluation status of a theory; that is, cumulation is measured in terms of theoretical growth, not the findings per se (Webster and Kervin 1971; Willer 1987). One consequence of this perspective on cumulation is that it encourages efforts to broaden and extend theories more than extensive empirical evaluation.

Empirical Experimentation

In contrast to theoretical experimentation, empirical experimentation assumes that research should provide broad coverage of some phenomenon, rather than focus in depth on a few theoretical principles. By broad coverage is meant that empirical experimenters try to identify as many factors that may impact a process or relationship and systematically investigate as many of these as time or a given experimental paradigm permits (see, e.g., Secord 1977). This holistic approach tends to treat each factor as if equal in importance; interrelationships among the factors tend to emerge empirically through statistical analysis rather than theoretically.[9]

In line with the strategy of broad investigation, empirical experimentation relies more heavily than theoretical experimentation on implicit substantive metatheory and intuition to generate research questions and to make corresponding procedural decisions. This is significant for two interrelated reasons. First, a given metatheory may be the source of several theories, each of which may predict different behaviors and have different domains. To the degree that empirical experimentation does not produce an explicit theory with a clearly defined set of scope conditions, but rather implicit theorizing and hypothesizing, it becomes easier to overlook implied scope conditions associated with a particular laboratory setting. For instance, some researchers group all experimental games together, ignoring the fact that subjects may perceive significant differences between a prisoner's dilemma and a chicken structure matrix (Schlenker and Bonoma 1978). While prisoner's dilemma and chicken structures are both dichotomous choice matrix games, making a self-interested choice (competition) in the prisoner's dilemma always guarantees a better outcome than making a collectively-interested choice (cooperation); no such guarantee holds for chicken structures, which are often described as "dangerous" games (Schelling 1960). Similarly, researchers using the Deutsch and Krauss trucking game often seem to overlook significant structural differences that may impact subjects' definition of the situation. In this locomotion game, subjects represent trucking firms that move goods to a destination point; each must choose between moving goods along a direct, profitable route that is common to both firms or a longer alternate route that can be used independently by each firm (Deutsch and Krauss 1960, 1962; Krauss and Deutsch 1966). Several studies using this game neglect differences between explicit and tacit bargaining when they manipulate the availability of communication channels so that subjects can discuss ways to resolve conflict over the short route (Deutsch 1973). Thus, when researchers fail to recognize implicit statements of conditionalization, they are more likely to make unwarranted comparisons between findings produced in different research settings.

Second, implicit metatheorizing without theory can raise questions about the methodological adequacy of a given test. This may occur because it is difficult, if not impossible, to work backwards from the features of a given setting to infer a researcher's goals and intentions or assumptions about the setting (Nemeth 1972; Tedeschi, Bonoma, and Brown 1971). Implicit theorizing can also impact the precision with which key variables are operationalized. Obviously, this too can have implications for assessing and interpreting findings, especially those that are inconsistent with other studies.

This issue seems to be at least partially involved in Nemeth's (1972) critique of research utilizing an abstract prisoner's dilemma matrix to study bargaining processes. In this critique, Nemeth challenges the fit between theoretical properties generally associated with bargaining, the way the matrix is ostensibly interpreted by subjects as well as researchers' implicit assumptions about actors' desire to maximize outcomes. Nemeth concludes that failure to articulate underlying assumptions and to critically examine the supposed fit between bargaining contexts and matrices used in the laboratory has produced findings that are difficult to interpret and apply to bargaining settings. Problematic findings have not produced theory but rather an emphasis on methodological investigations of the matrix game itself (e.g., using money vs. points for incentives; using red vs. black to denote the competitive choice).

Empirical experimentation also differs from theoretical experimentation with regard to the way findings are used. While theoretical experimenters use findings to help assess the status of a given version of a theory, empirical researchers frequently try to link their findings directly to contexts outside the laboratory. For this reason, many empirically-oriented experimenters tend to be preoccupied with external and ecological validity (Bushell and Burgess 1969; Pruitt and Kimmel 1977; Schlenker and Bonoma 1978; Webster and Kervin 1971). To address concerns about external or ecological validity, empirical experimenters advocate several procedural strategies, including use of field experiments, enhancing mundane realism in the laboratory (i.e., making the laboratory more like the world outside), systematically varying factors that might represent experimental limitations (e.g., subject characteristics, instructional variation, demand characteristics), and increasing the number of ways that a variable may be measured (Aronson and Carlsmith 1968; Rapoport 1966; Schlenker and Bonoma 1978). The last three strategies imply that to some degree, empirical experimenters adopt a "cup half empty" perspective on the issue of the highly standardized, artificial laboratory setting, and suggest that generalizability of findings (external or ecological validity) may be enhanced by showing how subject characteristics or context characteristics interact with independent variables (Aronson, Brewer, and Carlsmith 1985; Pruitt and Kimmel 1977).

In summary, experiments may be used to test and build theory, or they may be used empirically to provide broad coverage of a phenomenon. These two strategies are based on divergent assumptions about the purpose of research, the meaning of an experimental setting, and how findings produced in the setting should be treated. These differences impact concrete procedural decisions in several ways. First, experiments designed to test and build theory explicitly connect more paradigm features to metatheoretical and theoretical variables than those designed to test hypotheses or explore a phenomenon. An explicated metatheory and theory provide substantial guidance about the kind of experiment to be done, how the setting should be constructed, what variables should be controlled and in what way. Experimental limitations are recognized, but unless they can be linked to theory, they will not be introduced as independent variables. By directing which variables are to be manipulated and which are to be controlled, theory and metatheory dictate which data are produced and interpreted. From this perspective, being able to manipulate a large number of variables is of little consequence if those variables do not have immediate theoretical significance.

Second, theoretical experimenters regard the laboratory setting as an artificial creation that empirically represents conditions and concepts specified by a theory, not as an analog of anything outside the laboratory. This implies that laboratory features correspond to theoretical features, no matter how simple a theory may be, or how many "real life" features it ignores. Third, in theoretical experimentation, the adequacy of findings is not measured in terms of whether they tell the theorist anything about events, persons, or situations outside the laboratory, but in terms of whether they contribute to theory evaluation and cumulation. This suggests that the quest for findings high in "external validity" (see Campbell and Stanley 1966) is misplaced. Overall, these properties of theoretical experimentation are significant because they explicitly acknowledge the inseparability of theory, data, and the methods that produce data.

In the next section, I briefly outline five orienting metatheoretical assumptions that have provided the backdrop for developing bilateral deterrence and conflict spiral theory. Next, I note how these assumptions simultaneously serve an orienting function in making procedural decisions in the laboratory. Thus, this discussion provides the backdrop for analyzing the linkages between the program's metatheory, theory, and basic laboratory paradigm features, including control decisions.

SUBSTANTIVE PUNITIVE POWER METATHEORETICAL ASSUMPTIONS: IMPLICATIONS FOR PROCEDURAL DECISIONS

As described earlier, the power processes program offers an abstract theoretical position regarding the link between the power capabilities of two parties in conflict and their use of that capability during explicit bargaining (for a more complete description of the program's assumptions, see Bacharach and Lawler 1981; Lawler 1992; Lawler and Ford 1993). The program's orientation to this question is substantively framed by basic assumptions about the focal situation, the actor, and relevant social action. These have simultaneously shaped experimental procedural decisions.[10]

Basic Program Assumptions

Theorizing in the program is guided by five basic assumptions. First, social structure is the primary source of conflicting interests, implying that divergent interests stem from one's position in a social structure rather than from differences in personal preferences. Because social structures persist over time, conflicts linked to them resist resolution and are likely to recur regardless of the particular occupant of a position. Second, incentives that stimulate bargaining are also embedded in social structures that differentially allocate power and status over a range of social positions. Social structural conflicts tend to create incentives for actors to engage in distributive bargaining, which implies that actors use their power and status to gain advantage over each other. Third, actors have incomplete information and have incentives to manipulate the other's cognitions and withhold information to gain advantage. This implies a "boundedly rational" actor who can take action that departs from strict economic rationality. Fourth, conflict makes power salient and activates tactical action in bargaining. Tactics refer to moves or sets of moves that are directed at overcoming another's resistance, are grounded in the power relationship, and may be conciliatory or hostile in nature. Power is conceptualized in nonzero terms, meaning that one actor's power is not related to the other's in any a priori way; either both actors may increase or decrease in power. It is assumed that power capability should be distinguished from its use and outcome. This means that a power capability may or may not be used, and if used, has an uncertain probability of success. Fifth, bargaining consists of tactics and countertactics by both actors that have emergent effects on bargaining outcomes. This implies that, while social structure frames actors' definitions of the context, it does not fully determine their tactical behavior. For example, lower power parties may behave in ways that are inconsistent with their disadvantaged position and provide more resistance than a higher power actor would anticipate (Lawler and Ford 1993). Together these assumptions establish a set of core theoretical problems for bilateral deterrence and conflict spiral theory that focus on variations in power relationships (e.g., power equality vs. power inequality) and its impact on tactical action and outcomes in bargaining (e.g., whether power use is actually effective or not).

Beyond shaping theory construction, these very broad, basic assumptions impact concrete procedural decisions in two basic ways. First, because they present a particular image of social reality through rather broad assumptions about the focal situation, the actor, and relevant social action, they serve an orienting function in designing an experiment and concretizing these ideas into a plausible setting that will allow introduction of independent variables and measurement of dependent variables. For example, the emphasis on bargaining as a form of social action suggests an event (behavioral) rather than a judgement (perceptual) experiment. Specifically, the emphasis on bargaining, and tactical use of power as a response to conflict implies that the experimental setting will involve actual bargaining interaction between subjects, as opposed to simply describing a situation to subjects and assessing their perceptions, attitudes, or opinions about the situation.[11]

The metatheory offers a minimalist perspective of the actor; actors are described only in terms of their positions or roles in a social structure and in terms of some simple assumptions about how power is interpreted by the actor (Lawler, Ridgeway, and Markovsky 1993). This suggests two things. First, the metatheory does not entirely omit consideration of some internal, interpretative states. To assess these states, some kinds of attitudinal measures will be needed, although behavioral measures will be given priority. Second, the cover story (the study rationale as described to the subject) will describe a context in which actors (subjects) are placed in a theoretically relevant role or social position, for example, group representative. This role need not be described in great detail, but it should provide some information about conflict and power in the situation and the kinds of (tactical) choices that are available to the actors.

The emphasis on bounded rationality of actors, in combination with the emphasis on hostile and conciliatory tactics that have uncertain outcomes, suggests something about concrete features of the bargaining task. Specifically, it is important that the task create some sense of ambiguity for subjects and allow them to choose and respond to hostile and conciliatory actions in several ways. One way to create ambiguity for subjects is to restrict information about the other, for example, possible outcomes or stakes in the conflict.

The emphasis on explicit bargaining also provides some guidance about constructing the bargaining task. A defining characteristic of explicit bargaining is that it is formally recognized and that conciliatory tactics (e.g., offers and agreements) are clearly distinguishable from hostile tactics (threats and damage). Further, threatening behaviors are clearly distinguishable from damaging behaviors (the former communicates an "if-then" contingency; the latter actually reduces the other's outcomes). This implies that the subjects should be able to interact and communicate through offers, agreements (or nonagreements), threats, and damage. Finally, the nonzero sum approach to punitive power suggests that operationalizations of punitive power should truly reflect the range of variation that may occur in a relationship; one way to do this is to operationalize punitive power in terms of the percentage of damage that one subject can do to the other.

Second, because they provide information about which features of social reality are important, the metatheoretical assumptions also have implications for implementing experimental control of theoretically relevant and theoretically irrelevant variables. For example, from the perspective of a given metatheory, some variables that are not explicitly included in a theory may have theoretical potential while others will remain theoretically uninteresting; response latency may have theoretical potential for metatheories concerned with the process of learning but not for metatheories concerned with power processes. I return to the theme of metatheory and experimental procedures in the last section of this paper where I describe the features of the punitive power setting in more detail and show how both theory and metatheory shape these decisions. However, before I do this, I discuss the theoretical development of bilateral deterrence and conflict spiral theories in more detail and briefly describe empirical tests of these theories. A more complete description of the friendly competition between these two theories makes it easier to understand the experimental setting designed to test them.

BILATERAL DETERRENCE AND CONFLICT SPIRAL THEORIES: USING EXPERIMENTS TO FACILITATE FRIENDLY COMPETITION

The punitive power branch of theory and research represents an effort to resolve apparent contradictions on the effects of punitive power in bargaining suggested by Tedeschi, Schlenker, and Bonoma's (1973) classic work on threats and influence vs. Deutsch and Krauss's (1960, 1962) seminal work on conflict escalation. The former showed that increases in threat magnitude and credibility increased compliance in a unilateral power situation, while the latter showed that punitive capabilities were tempting to use and that use engendered resistance (to avoid a loss of face) rather than compliance (see Bacharach and Lawler 1981; Lawler 1986, 1992; Lawler and Ford 1993).

From these contradictory positions, Bacharach and Lawler (1981) and Lawler (1986) developed two initial formulations, now termed *bilateral deterrence* and *conflict spiral*. Both theories predict the use of punitive tactics, both apply to explicit bargaining contexts, and both have undergone five stages of development by utilizing the program's explicated metatheory and the strategy of friendly competition.

A first stage formalized the two classic arguments to sharpen the contrast between the two approaches and enable further analysis (Bacharach and Lawler 1981). Second, additional analysis made clear that the formulations interpreted the meaning of power and the basis of its use differently. The deterrence formulation interpreted power in terms of fear of retaliation and traced A's use of power to B's power level; the conflict spiral formulation interpreted power in terms of temptation and traced A's use of power to A's power level (Bacharach and Lawler 1981).

At this point, neither theory took account of an actor's assessment of the other's willingness to use punitive tactics. By incorporating an additional concept, *expectation of attack*, both theories more clearly predicted variations in total power. Bilateral deterrence theory now predicted that increases in total power would decrease punitive tactic use; conflict spiral predicted the reverse.

A third stage of development resolved theoretical ambiguity concerning equal vs. unequal power predictions by positing a difference in interpretation between the two types of power relationships. In equal power relationships, actors would weigh the intervening cognitive variables equally; they would weigh them differently in unequal power relationships. Bilateral deterrence theory now specified that higher power actors would pay more attention to their reduced fear of retaliation; lower power actors would pay more attention to an increased expectation of attack. This would produce more use by both actors.

Conflict spiral now predicted that higher power actors would stress their reduced expectation of attack; while lower power actors would pay more attention to their own reduced temptation to use power. This would produce less use by both actors. Thus, bilateral deterrence theory now predicted more usage when parties have unequal as opposed to equal power, while conflict spiral theory predicted the reverse (Lawler 1986).

A fourth stage developed a conditionalization principle and an extension to concession tactics. The conditionalization principle specified conditions under which deterrence vs. spiral effects are likely to occur. Lawler, Ford, and Blegen (1988) theorized that conditions making fear of retaliation salient would produce deterrence effects, while conditions making temptation salient would produce conflict spiral effects. The extension to concession tactics is based on the assumption that punitive tactics are negatively related to concession tactics—thus, bilateral deterrence would predict higher rates of yielding and agreements when actors have higher rather than lower total power and when actors have equal rather than unequal power. In contrast, conflict spiral would predict the opposite.

A fifth stage of development entailed empirical evaluation of the four issues emerging from theoretical friendly competition. These included the total power prediction, the relative power prediction, the concession generalization prediction, and the conditionalization prediction. In accord with the spirit of friendly competition, two experiments pitted bilateral deterrence and conflict spiral theories against one another to assess the first three predictions (Lawler, Ford, and Blegen 1988). A third experiment utilizing a programmed other examined issues of conditionalization for bilateral deterrence theory by varying the probability that a bargaining opponent would initiate an unprovoked attack and the probability that the opponent would retaliate for the subject's use of punitive tactics in an unequal power context. Thus, this experiment closely examined the intervening variables of fear of retaliation and expectation of attack and their links to power use for unequal power actors (Ford and Blegen 1992).

Results from the first two experiments generally corroborated bilateral deterrence predictions over those of conflict spiral (Lawler, Ford, and Blegen 1988). In the later stages of bargaining, higher levels of total power were associated with less punitive tactic use, and more punitive tactics were used in unequal than in equal power relationships, suggesting that actors needed to experience the costs of using punitive tactics before acting in accord with deterrence. As predicted, no differences emerged between high and low power actors' use of punitive tactics. Similar (though weaker) effects were obtained for concession tactics. Larger concessions were made by those in high total power relationships relative to those in low total power relationships, and those in equal power relationships conceded more than those in unequal power relationships. No significant effects were found for agreements.

Results from the third experiment also corroborated bilateral deterrence predictions (Ford and Blegen 1992). Higher probabilities of attack led to more use of punitive tactics; higher probabilities of retaliation led to less use of punitive tactics; higher probabilities of retaliation in combination with lower probabilities of attack produced the lowest rate of punitive tactic use. While low power actors used fewer punitive tactics overall than did high power actors, there were no differences in the way high and low power actors used punitive tactics; that is, low power actors were no more likely to use tactics to retaliate or to initiate unprovoked attacks than were high power actors. This pattern of findings raises the question of conditions under which low power actors would behave differently from high power actors. Finally, no differences emerged for concessions and agreements. In sum, the results of the three experiments provide substantial support for the basic predictions of bilateral deterrence theory over those of conflict spiral; the first two studies direct attention to power processes over time, while the third directs attention to incipient differences between high and low power actors. However, it is important to note that the evidence supporting bilateral deterrence theory highlights the need to locate conditions that will elicit conflict spiral processes; evidence on high and low power actors suggests the need for further theoretical clarification of power difference relationships.

Overall, the program relied less on empirical evaluation than on theoretical revision and extension. However, evaluation was necessary at key points to assess basic theoretical predictions and to provide further impetus to additional growth. The metatheory and theory established four primary problems: total power, relative power, concession behavior, and conditionalization. These problems, along with the metatheory and theories defined the features of the laboratory setting used to evaluate the theories. In the next section, these linkages are described in more detail.

METATHEORY, THEORY, AND EXPERIMENTAL CONTROL

As noted earlier, regardless of a researcher's general metamethodological orientation, the principal reason for using laboratory experimentation is that it provides a potentially high degree of control over theoretically relevant and irrelevant variables that enter into a setting, thereby allowing researchers to focus on specific issues at will. For instance, experimental control facilitates investigation into total and relative power, concessions, and conditionalizing bilateral deterrence and conflict spiral theories.

There are various ways of achieving control, for example, random assignment to conditions, holding variables constant, setting a specific range of values for a variable, or classifying them as irrelevant (Aronson, Brewer, and Carlsmith 1985; Aronson and Carlsmith 1968; Foschi 1980). In general, random assignment eliminates the effects of variables not included in a theory (or hypothesis) but which could impact results; holding variables constant controls variables that are theoretically or hypothetically relevant; specifying a range of values controls variables that are theoretically relevant, or that are not included in a theory, but which may impact results. In some cases, variables that are held constant or that have a limited range relate to the generality or scope conditions of the theory; controlling these variables directly bears on the confirmation status of the theory and provides information about the theory's predictions (Foschi 1980; Webster and Kervin 1971). In other cases, variables that are controlled through limiting their values may represent experimental constraints for a given study, for example, age, sex, educational composition of the subject pool, or specific features of the experimental setting (Foschi 1980). These variables may have theoretical or explanatory potential—if they can be abstractly conceptualized and linked to the theory, they may play an important role in theoretical growth (Foschi 1980). Clearly then, strategies for control are linked to the metatheoretical (e.g., social structural foundation of conflict) and theoretical status of variables (e.g., use of punitive and conciliatory tactics); as the status of specific variables shifts, so does the manner in which they are controlled.

Consider the relationship of metatheory and theory to using these forms of control and to creating the standardized context (including the "cover story") in a typical punitive power study. Recall that metatheoretical concerns dictate an impact experiment that allows observation and measurement of actual bargaining behavior, that subjects be placed in a social position like group representative that gives them access to power and clearly distinguishable tactical choices, and that they should confront some ambiguity and uncertainty. To concretize these features, the standard setting requires that two to six same-sex subjects be seated individually or in pairs in separate cubicles. Once in the cubicles, they are randomly assigned to read instructions which provide the "cover story" for the study, describe the task, and introduce the key independent variables. The instructions indicate that the subject is participating in a bargaining study and that, to preserve anonymity, each will be assigned to bargain with another person at random. One member of each pair represents a group identified as Alpha, the other represents a group identified as Beta; these groups are in a buyer-seller conflict over an issue or a commodity. Ostensibly, Alpha and Beta have had a preliminary discussion in which it is established that they disagree over an issue, and the subject's task is to represent their respective group during negotiations and take steps to maximize their group's profit regardless of the opponent's outcomes. This establishes an individualistic (as opposed to a cooperative or competitive) orientation during bargaining which presumably heightens the distributive features of the context. The issue to be negotiated may be represented very abstractly in terms of a difference between two numerical anchors or more concretely as the difference between two prices for a commodity. Payoff from the bargaining is represented in terms of points which are converted into money at the end of the study. Subjects receive no information about the other's profit or payoff at any point during the bargaining.

The subjects are also told that they have two sources of profit; one source refers to points that can be gained through negotiation. These points are affected only by the concession choices of each actor and are not subject to attack by the opponent. The subjects are informed that, if they fail to reach agreement, neither party earns points from the negotiations. The instructions also indicate that both bargainers begin the bargaining with a set number of points (described as an "account") which are vulnerable to attack by the opponent. The amount of damage, expressed as a percentage, that each actor can impose on the other is determined by experimental condition and represents the manipulation of punitive capability (the primary independent variable of interest). If subjects reach agreement, their payoff consists of the points associated with a given agreement level and the number of points remaining in their account. If no agreement is reached, their payoff consists solely of the points remaining in their account.

The instructions describe the bargaining task in terms of a sequence of offers, counteroffers, messages, and countermessages, termed a bargaining "round," which establishes a context for interaction. Subjects are told that there is a maximum number of bargaining rounds (e.g., 15) and that, if they do not reach agreement by this maximum, the bargaining will be terminated. In most of the studies, the offers and messages are entered on a computer terminal and transmitted to the other bargainer (although some studies have used handwritten offers and counteroffers, see Bacharach and Lawler 1981). The instructions tell subjects that on each round they have three offer options: they may repeat their previous offer, make a concession, or accept the other's last offer. The specific numbers used for the offer-counteroffer sequence cannot fall outside the issue continuum described in the instructions, nor can subjects concede along a dimension other than the one described. After the exchange of offers, subjects choose one of three message options. They may send a warning (threat), a fine (damage), or a "no message" to the other. Warnings indicate that a bargainer intends to levy a fine in the future if the other fails to make more concessions; warnings can be sent on any or all of the bargaining rounds but do not have to precede a fine. Fine messages actually reduce the other's account points by a fixed amount determined by experimental condition (e.g., 10%, 50%, 90%, etc.). Both subjects know their own and the other's damage capability. The no-message option represents a neutral communication subjects send if they do not want to send a warning or a fine. After the exchange of messages, a new round begins. The offer-counteroffer (concessions) and message-countermessage sequence (no message, warning, fine) along with rates of agreement represent the core dependent variables for the studies.

Finally, the instructions indicate that when the bargaining ends, the subjects will be asked to complete a series of questions that relate to their perceptions of the bargaining. These questions are presented in a likert-type format and are designed to gather information on the intervening variables of fear of retaliation, temptation, and expectation of attack. Fear of retaliation is generally measured by asking subjects if they were concerned that the other would retaliate with a fine if they used one first. Expectation of attack is measured by asking subjects if they thought the other would use fines regardless of whether they did. Temptation is measured by asking subjects if they intended to use fines regardless of whether the other did.

While some of these features are undoubtedly accounted for by ad hoc decisions, many are guided by metatheoretical and theoretical concerns that specify the kind of problems the setting must investigate and how to view variables that enter into the setting. As described above, friendly competition between bilateral deterrence and conflict spiral has produced four core theoretical problems (variations in total power; variations in relative power; extension to concessions, conditionalization) which have had implications for procedural decisions in the laboratory.

First, as in most experiments, subjects are assigned randomly to conditions in order to preclude personal characteristics like personality differences, motivational factors, and learning differences from having a differential impact on experimental conditions. Such psychologically-based individual differences are not of interest to the metatheory guiding the program of research and will likely continue to be dealt with through random assignment. However, some characteristics of subjects, for example, gender, educational level, and race, may be conceptualized in terms of their structural properties, like status. This implies that while the theories themselves do not predict a gender, educational, or racial difference or account for how these might influence use of punitive tactics, these variables should be dealt with by holding them constant (e.g., same-sex dyads) or allowing a narrow range of variation. Some studies have counterbalanced the number of all-male and all-female dyads in each condition and have found no difference between the rate at which males and females use punitive tactics (see e.g, Lawler, Ford, and Blegen 1988). If this evidence continues to mount, gender may eventually be regarded as an irrelevant variable (if the dyads are same sex). However, if differences are detected, then efforts will be made to abstractly conceptualize gender to theoretically account for such differences.

Separating subjects from one another serves three important functions. First, it eliminates the possibility that differences in status, demeanor, dominance, or other influence-related behaviors not currently accounted for by the theories could impact results. Second, it partly controls measurement of punitive tactics. For instance, threats may be communicated explicitly or tacitly; the former are made in direct statements of intent to harm while the latter are made indirectly through nonverbal movements—for example, glares, frowns, and other facial movements or through verbal statements that do not clearly communicate an "if-then" contingency and that may not be well distinguished from the application of damage. Subjects may well vary in the extent to which they rely on various means of communicating threats as well as how they interpret various kinds of communications. This feature is important to both the metatheory and the theories themselves; the metatheory suggests the importance of both threats and damage, while the theories only predict the *use* of power to damage outcomes. Third, the program's metatheory assumes that ambiguity and uncertainty are key features of conflict situations; preventing face-to-face discussions serves to heighten the sense of uncertainty as neither subject can discuss strategy with the other. Thus, some of the variables controlled by the current nonface-to-face setting represent subtle metatheoretical emphases, others represent matters of scope, while still others represent experimental limitations. By preventing subjects from using nonverbal threats, the setting creates conditions relevant to the scope of the theories in the program; by enhancing ambiguity, the setting instantiates a core assumption about conflict situations; controlling the form of communication represents an experimental limitation of the setting (e.g., the scope conditions do not indicate that the processes hold only when subjects cannot engage in face-to-face interaction).

Preventing subjects from knowing the identity of their bargaining opponent controls the possibility that preconceived ideas about the opponent might influence their bargaining behavior (e.g., attractiveness, friendliness, etc.). If preconceived ideas about the subjects as individuals were allowed to enter the setting, the subjects might not give as much attention to their role as a group representative for Alpha or Beta. The minimal intergroup character of the situation is important to the metatheory for two reasons. First, the metatheory indicates that conflict and interests are linked to social positions rather than personal preferences. Second, the metatheory suggests the importance of isolating different kinds of power that might be operating in a bargaining situation. The bilateral monopoly situation between Alpha and Beta serves to isolate punitive power processes from dependence power processes by holding dependence high and constant; if either or both subjects have alternative bargaining opponents, dependence power could be confounded with punitive power. In addition, the bilateral monopoly situation is important to both theories because they relate power use to the power relationship between two parties. Thus, the bilateral monopoly is part of the operationalization of one of the scope conditions for the theories.

By informing subjects that they have two sources of profit, one affected by concessions, the other by damage tactics, the setting allows clear and separate measurement of conciliatory and hostile behavior. Presumably translating the profit points into money heightens a sense of conflict and provides a strong incentive to bargain—both of these features are relevant to the substantive metatheory which stresses that conflict makes power and tactical assessment salient to bargainers and that the incentive to bargain is embedded in a social structure. While early versions of the theories predicted only punitive tactics and did not include a specific prediction for concession tactics, the metatheory has always stressed the importance of distinguishing between hostile and conciliatory tactics. Thus, being able to clearly separate the two forms of tactics has had implications for assessing the theories' predictions.

The bargaining task developed for this setting operationalizes the primary scope condition of explicit bargaining. All subjects interact in the same manner and have available to them the same offer and message options from which to choose and have the same temporal restrictions applied to them. By structuring interaction into a sequence of offer-counteroffer and message-countermessage, the task minimally instantiates the open communication process characteristic of explicit bargaining. A second feature of explicit bargaining is instantiated by providing subjects with an issue that has a range of possible solutions. A third feature of explicit bargaining is created by allowing subjects to choose not to agree; this gives their offers the provisional character of offers in explicit bargaining.

In addition, by providing subjects with several options from which to choose (e.g., concede, stand firm, agree; send threats, damage, or neutral communication), the setting allows for actors' latitude during conflict situations. This means that the subjects can choose to engage in supposedly "nonrational" forms of behavior like nonagreement. Of the three message options, fines are the most important, as their value represents the implementation of the key independent variable in the setting, and their use represents the key dependent variable predicted by the theories of punitive power.[12] While subjects can make as many or as few concessions as they like, the number of fines they may send is generally restricted. This facilitates control over the size of each fine and the overall magnitude of damage that each can levy on the other, thereby permitting a more precise test of the theories' basic predictions about punitive tactic use.

By standardizing messages for subjects, their meaning is relatively clear—the meaning of warnings is not confused with that of fines, and subjects are not free to improvise and create their own messages. Third, the offer-message sequence over rounds allows analysis of bargaining behavior over time—for example, initial offers, final offers, overall concession magnitude (the difference between first and last offers), trends over the bargaining rounds, and agreement rates while controlling for the form and type of concession, which permits assessment of the theoretical extension to concession behavior. Similar kinds of analyses can be performed for messages (e.g., overall rates, trends over time, impact on agreements, and relationship to concessions). Additionally, patterns of use for all three messages allows deeper insight into the interrelationships among potential, use, and outcomes of a power process, a core metatheoretical concern. For instance, a party with a large punitive capability facing a much lower power opponent, and who uses a large number of warnings or no messages relative to fines, may be able to influence without actually using the power capability; another party may decide to resist influence attempts by an opponent who uses a large number of fines despite the prospect of poor outcomes from the bargaining. The standardization of messages also means that the basic setting can be modified to examine how programmed patterns of tactical action impact subjects' use of punitive tactics, for example, a programmed other permitted examination implications of the theories' intervening variables by systematically varying the probability that the programmed other would retaliate for hostile acts and initiate unprovoked hostile acts (Ford and Blegen 1992).

Real negotiations vary substantially with respect to time frame and the actual number of offers and messages. Some research has established that time affects various bargaining behaviors, for example, frequency of conceding, probability of agreement, and punitive tactic use (see, e.g., Rubin and Brown 1975). Controlling time makes measurement of both conciliatory and hostile tactics easier and eliminates the possibility that differing amounts of time affect results. The significance of time as a mediating factor cannot be overlooked, as empirical evidence suggests that some key theoretical processes (e.g., deterrence effects) may emerge over time as actors have the opportunity to directly experience the impact of punitive tactic use (Lawler, Ford, and Blegen 1988). Such results suggest the importance of reviewing the metatheory for guidance in considering the role of time. As the program moves toward consideration of power change, rather than power use tactics, time will likely assume an even greater theoretical role (see Lawler and Ford 1993).

The post-experimental questionnaires help to gather data relevant to both metatheoretical and theoretical concerns. As noted earlier, the power processes metatheory stresses the importance of both action and subjective interpretation, while conveying a tilt toward action. Specifically, power processes are presumed to be mediated by cognitive, subjective processes. For this reason, subjects are asked a series of questions relating to the cognitive variables specified by the theories (expectation of attack, temptation, and fear of retaliation). These questions allow examination of key theoretical propositions, for example, larger levels of power in an equal power context produce higher levels of fear of retaliation by both parties. Other questions assess reasons for using punitive tactics and perceptions of concessions. These questions give some insight into what actors think they accomplish when they use fines or make concessions. Finally, some questions are directed at assessing whether subjects perceived themselves to be in conflict with another; this represents an important scope condition for the theories. While subjects are not necessarily expected to uniformly indicate that they perceived a *high* degree of conflict, those indicating that they did *not* perceive a conflict did not try to get a good agreement for their party or who worked for the benefit of both would be eliminated from analysis.

Finally, the "friendly" competition between bilateral deterrence and conflict spiral theories has meant that the same setting can be used to contrast the theories' contradictory predictions because the theories *share the same set of scope conditions and predict the same dependent variables*. This means that empirical cumulation has benefitted from the strategy of parallel development between the two theories. For example, because of the conceptual overlap between the two theories, general procedures for testing the theories' basic predictions for power use and concessions do not differ. However, efforts at conditionalization, that is, understanding factors that increase the salience of temptation vs. fear of retaliation, may require special modifications of the setting that focus on one theory at a time (e.g., see Ford and Blegen 1992). Further, if such efforts result in a change in the scope conditions for one of the theories, a different setting will have to be used to test additional predictions for that theory.

As this analysis reveals, even highly simplified standardized settings involve a number of complex procedural decisions, some of which may be ad hoc, but many of which are connected in important ways to the problem focus of the program, its supporting substantive assumptions, and theories. Some metatheoretical considerations deal with substantive concepts and assumptions in a program, while others deal with general strategies for theoretical development. Isolating dependence from punitive power is an example of the former. The latter is illustrated by the use of one standardized setting to contrast the contradictory predictions of bilateral deterrence and conflict spiral theory.

CONCLUSION

This paper provides a metamethodological analysis of the experimental setting used in a program of theoretical research on punitive power in bargaining by discussing the ways that explicated metatheory and theory impact procedural decisions in the laboratory. In addition, the paper discusses the role of the standardized setting in testing predictions of theoretical variants sharing the same scope conditions, that is, bilateral deterrence and conflict spiral theory by showing how experimental control is used procedurally to create conditions relevant to testing both theories.

This kind of analysis is important because many experiments fail to explicate assumptions that guide construction of a setting's concrete features and to show how these features link to the more abstract ideas the setting presumably helps to evaluate. Failure to precisely articulate underlying metatheory and theory increases interpretative and reliability difficulties, makes it more likely that implicit statements of conditionalization will be ignored, and detracts from efforts to improve theoretical growth and revision.

As part of my analysis of the link between metatheory, theory, and experimental procedures, I contrasted an empirical approach to experimentation with a theoretical one. These strategies have much in common—both select experimentation because it affords a high degree of control and helps to eliminate more extraneous variables and alternative hypotheses than many other types of methodology. Both strategies also require many of the same types of decisions—for example, type of experiment, how to operationalize independent variables, how to measure dependent variables, and others.

Nevertheless, the two strategies differ in key ways. The empirical strategy is used for broad investigation of a phenomenon and is guided by implicit metatheory and concrete hypotheses. One important metamethodological assumption characteristic of this approach is that more data are better than less because they provide a more complete picture of the phenomenon under investigation and that better theories eventually result from better understanding of the phenomenon (i.e., more data on the subject). To some extent then, an empirical approach tacitly assumes that the "data speak for themselves" and that experimenters can conduct experiments to "see what will happen" (see Webster 1994). This also suggests that empirical experimenters implicitly act as if a situation they create can somehow take on a "life of its own" and produce data that transcend limitations inherent in the setting. Consequently, empirical experimentation is also more likely than theoretical experimentation to treat the laboratory setting as an analog for the phenomenon or process under investigation and to entail changes in the setting without considering how such changes might implicitly alter the scope or comparability of findings.

Theoretical experimentation is most concerned with developing implications of several scope constrained theoretical principles to produce theoretical cumulation and with using an explicated metatheory along with the theoretical principles to develop a highly simplified setting to create and recreate theoretically relevant conditions for purposes of both theory assessment and theory revision. Theoretical experimentation does not link the laboratory setting to the world outside the laboratory but rather uses the theory it tests to form the link (Wagner 1984; Webster and Kervin 1971).

Epistemological differences between these strategies suggest that procedural decisions may be viewed as primarily ad hoc, guided by implicit metatheoretical or implicit theoretical assumptions, or driven by explicated metatheory and theory. Because laboratory experiments are by definition artificial, highly unusual contexts produced by a researcher, they are inevitably guided by a host of assumptions about the nature of social reality, properties of the actor, and the situation (see Webster 1994). To the degree that underlying assumptions are articulated, researchers can provide better tests for their ideas and more adequately assess the relevance of a particular set of data for a theory or metatheory.

This paper provides an analysis of the way that the power processes program's explicated metatheory and formalized theory shape procedural decisions in the laboratory and so increase the probability of further theoretical growth. Among the metatheoretical assumptions that play a role in such decisions are the idea that conflict has a social structural foundation, that power is related to social structure, that actors may bargain to resolve conflict, and that their power capabilities become salient in the presence of conflict. Also important are ideas about the way power should be conceptualized (e.g., in nonzero sum terms), and as a process with three distinct phases (e.g., capability, use, and outcome). Use is further conceptualized in terms of two broad categories of tactics: hostile vs. conciliatory.

By encouraging precise specification of key concepts in bilateral deterrence and conflict spiral theories, these directives also aid in operationalizing variables in the experimental setting and in making decisions about how to create and control features of the setting. More specific guidance comes from the theories themselves (i.e., scope conditions, independent and dependent variables, etc.). The more clearly the setting links to core metatheoretical concerns and theoretical variables, while controlling variables that are not theoretically relevant, the more adequate the test and the better the prospects for interpreting findings and improving cumulation over a series of studies.

The experimental setting designed to test power processes theories can broadly be described as an impact experiment that creates a bilateral monopoly, explicit bargaining context. This context allows subjects to openly exchange offers and counteroffers along an issue continuum with multiple solutions. This property is required by the scope conditions for both bilateral deterrence and conflict spiral theories. The major theoretical independent variable specified by both theories is the amount of punitive capability each party can levy on the other; the major dependent variable is the frequency with which this capability is used. To operationalize this, the setting provides both subjects with the ability to reduce the other subject's monetary outcomes by a fixed percent. While the primary dependent variable predicted by the theories is the frequency of punitive tactic use, the setting permits analysis of other variables that are theoretically relevant (e.g., concession and agreement behavior) as well as interrelationships between concessions and punitive behavior. While the setting does allow analysis of variables that are not explicitly predicted by the theories, it still closes out many processes (e.g., status, legitimation, justice, and emotional processes are eliminated from the setting). Such simplification is necessary to create the best possible test for the theories under their scope conditions.

Changes in the setting will emerge as friendly competition suggests other ways for the theories to grow; for example, the theories may develop an account for the role of processes currently closed out of the setting. Or it is possible that friendly competition may identify a scope condition that differentiates the domain of the two theories. Either of these theoretical modifications will require corresponding modification of the basic setting to evaluate theoretical predictions and stimulate additional theoretical growth. Overall, in contrast to empirical experimentation, theoretical experimentation pursues a distinctive strategy for building, and using, experimental settings, and for dealing with experimental limitations. Fundamentally, this strategy recognizes that metatheoretical and theoretical assumptions guiding a program of work inevitably produce their own data.

ACKNOWLEDGMENTS

I would like to thank Edward Lawler, David Wagner, Cathy Johnson, and most of all, Thomas Ford for their helpful comments and support.

NOTES

1.   Some of the research uses an experimental vignette method (see, for example, Bacharach and Lawler 1976, 1981). These studies describe a conflict between two parties (e.g., employer-employee), ask subjects to take the role of one of the parties and then report their perceptions of power for self and others, and indicate the likelihood of making various tactical choices. Power (dependence) is manipulated by varying the described stakes and alternatives for both parties. While subjects do not actually bargain with one another, the vignettes abstract the essential features of concern, for example, a conflict situation, non-zero sum power, and tactical options, that would occur in a laboratory setting.

2.   Efforts to capitalize on the contradictory elements by pushing parallel theoretical growth do not characterize true theoretical competitors as described by Wagner (1984, chapters 3-4). Wagner indicates that true competitors are imperialist in their quest to take over another theory's domain. What makes competition of this sort difficult to assess is that competitors (as opposed to variants) have dissimilar structures but similar problem foci. consequently, there is little agreement over basic conceptualization, predictions, and relevant evidence. Wagner (1984, p. 59) concludes that, in general, competition generates "much heat, but little light."

3.   For an example of three paradigms that have been analyzed theoretically, see Foschi (1980), Tedeschi, Bonoma, and Brown (1971), and Willer (1987).

4.   For an example of a somewhat different approach to creating laboratory procedures, see Aronson and Carlsmith (1968) and Aronson, Brewer, and Carlsmith (1985). These two papers provide a very detailed, pragmatic, although generic analysis of the process by which ideas may be experimentally tested in the laboratory and in the field.

5.   Similar procedural concerns appear in analyses of qualitative methodology; see, for example, Holstein and Gubrium (1995).

6.   Prisoner's dilemma situations are mixed motive, interdependence contexts in which two or more parties must choose between cooperation and competition. From the perspective of each actor, the best payoff is produced by choosing competition when the other actor chooses to cooperate; the worst payoff is produced by choosing cooperation when the other chooses competition. Mutual cooperation produces a moderately beneficial payoff for both; mutual competition produces a moderately costly payoff for both.

7.   See Rubin and Brown (1975) for a somewhat different perspective on the prisoner's dilemma matrix.

8.   Cover stories are the verbal or written statements administered to subjects that provide a pretext or rationale for the study. In some cases, the cover story contains information that represents manipulation of the independent variables.

9.   Interestingly, a similar point is made about qualitative field research by Holstein and Gubrium (1995), who suggest that many field researchers see themselves as "prospectors," who mine subjects for uncontaminated data.

10.   Berger, Wagner, and Zelditch (1992) provide a similar analysis of the status characteristics program.

11.   The cognitive and interpretative emphasis in the metatheory implies that vignette studies, which do describe a situation and assess subject opinions, may be useful if a given study focuses on interpretative issues. In fact, several early studies in the program used a vignette role-playing method to test hypotheses about power perception (see Bacharach and Lawler 1976).

12.   This effectively eliminates the possibility of logrolling, or trading off, on issues of lesser importance to either party. Multiple issues with differing priorities is a core feature of an integrative bargaining context (see, e.g., Pruitt 1982); the power processes program focuses on distributive contexts and so relies on single issue conflicts.

REFERENCES

Aronson, E., M. Brewer, and J. M. Carlsmith. 1985. Pp. 441-486 in *The Handbook of Social Psychology*, 3rd ed., Vol. 1, edited by G. Lindzey and E. Aronson. New York: Random House.

Aronson, E., and J. M. Carlsmith. 1968. "Experimentation in Social Psychology." Pp. 1-79 in *The Handbook of Social Psychology*, Vol. 2, edited by G. Lindzey and E. Aronson. Reading, MA: Addision-Wesley.

Bacharach, S., and E. J. Lawler. 1981. *Bargaining: Power,Tactics and Outcomes.* San Francisco: Jossey-Bass, Inc.

Berger, J. M., H. Fisek, R. Z. Norman, and M. Zelditch, Jr. 1977. *Status Characteristics and Social Interaction.* New York: Elsevier.

Berger, J. M., D. G. Wagner, and M. Zelditch. 1989. "Theory Growth, Social Processes, and Metatheory." Pp. 19-42 in *Theory Building in Sociology: Assessing Theoretical Cumulation*, edited by J. H. Turner. Newbury Park, CA: Sage.

_____. 1992. "A Working Strategy for Constructing Theories: Status Organizing Processes." Pp. 107-123 in *Metatheorizing*, edited by G. Ritzer. Newbury Park, CA: Sage.

Boyle, E. H., and E. J. Lawler. 1991. "Resolving Conflict Through Explicit Bargaining." *Social Forces* 69: 1183-1204.

Bushell, D., Jr., and R. L. Burgess. 1969. "Characteristics of the Experimental Analysis." Pp 27-48 in *Behavioral Sociology: The Experimental Analysis of Social Process*, edited by D. Bushell and R. L. Burgess: New York: Columbia University Press.

Campbell, D. T., and J. C. Stanley. 1966. *Experimental and Quasi-Experimental Designs for Research.* Chicago: Rand McNally.

Chertkoff, J. M., and J. K. Esser. 1976. "A Review of Experiments in Explicit Bargaining." *Journal of Experimental Social Psychology* 12: 464-486.

Cohen, B. P. 1980. "The Conditional Nature of Scientific Knowledge." Pp 71-110 in *Theoretical Methods in Sociology*, edited by L. Freese. Pittsburgh: University of Pittsburgh Press.

_____. 1989. *Developing Sociological Knowledge*, 2nd ed. Chicago: Nelson-Hall.

Deutsch, M. 1973. *The Resolution of Conflict: Constructive and Destructive Processes.* New Haven, CT: Yale University Press.

Deutsch, M., and R. M. Krauss. 1960. "The Effect of Threat on Interpersonal Bargaining." *Journal of Abnormal and Social Psychology* 61: 181-189.

_____. 1962. "Studies of Interpersonal Bargaining." *Journal of Conflict Resolution* 6: 52-76.

Emerson, R. M. 1962. "Power Dependence Relations."*American Sociological Review* 27: 31-40.

_____. 1972. "Exchange Theory, Part II: Exchange Relations, Exchange Networks, and Groups as Exchange Systems." Pp. 58-87 in *Sociological Theories in Progress*, Vol. 2, edited by J. Berger, M. Zelditch, Jr., and B. Anderson. Boston: Houghton Mifflin.

Ford, R., and M. A. Blegen. 1992. "Offensive and Defensive Use of Punitive Tactics in Explicit Bargaining." *Social Psychology Quarterly* 55: 351-362.

Foschi, M. 1980. "Theory, Experimentation, and Cross-Cultural Comparisons in Social Psychology." *Canadian Journal of Social Psychology* 5: 91-102.

Freese, L. 1980. "The Problem of Cumulative Knowledge." Pp. 13-69 in *Theoretical Methods in Sociology*, edited by L. Freese. Pittsburgh: University of Pittsburgh Press.

Gergen, K. J. 1973. "Social Psychology as History." *Journal of Personality and Social Psychology* 26: 309-320.

_____. 1978. "Experimentation in Social Psychology: A Reappraisal." *European Journal of Social Psychology* 8: 507-527.

_____. 1985. "The Social Constructivist Movement in Modern Psychology." *American Psychologist* 40: 266-275.

Gubrium, J. F. 1988. *Analyzing Field Reality.* Newbury Park, CA: Sage.

Henshel, R. L. 1980. "Toward the Deliberate Use of Unnatural Experimentation." Pp. 175-200 in *Theoretical Methods in Sociology*, edited by L. Freese. Pittsburgh: University of Pittsburgh Press.

Holstein, J. A., and J F. Gubrium. 1995. *The Active Interview*. Thousand Oaks, CA: Sage.

Krauss, R.M., and M. Deutsch. 1966. "Communication in Interpersonal Bargaining." *Journal of Personality and Social Psychology* 4: 572-577.

Lawler, E. J. 1986. "Bilateral Deterrence and Conflict Spiral: A Theoretical Analysis." Pp. 107-130 in *Advances in Group Processes*, Vol. 4, edited by E. J. Lawler. Greenwich, CT: JAI Press.

_____. 1992. "Power Processes in Bargaining." *Sociological Quarterly* 33: 17-34.

Lawler, E. J., and S. Bacharach. 1987. "Comparison of Dependence and Punitive Forms of Power." *Social Forces* 66: 446-462.

Lawler, E. J., and S. B. Bacharach. 1976. "Outcome Alternatives and Value as Criteria for Multistrategy Evaluations." *Journal of Personality and Social Psychology* 34: 885-894.

_____. 1986. "Power Dependence in Collective Bargaining." Pp. 191-212 in *Advances in Industrial and Labor Relations*, Vol. 3, edited by D. Lipsky and D. Lewin. Greenwich, CT: JAI Press.

Lawler, E. J., and R. Ford. 1993. "Metatheory and Friendly Competition in Theory Growth: The Case of Power Processes in Bargaining." Pp. 172-210 in *Theoretical Research Programs: Studies in Theory Growth*, edited by J. Berger and M. Zelditch, Jr. Stanford, CA: Stanford University Press.

_____. 1995. "Bargaining and Influence in Conflict Situations." Pp. 236-256 in *Sociological Perspectives on Social Psychology*, edited by K. S. Cook, G. A. Fine, and J. S. House. Boston: Allyn and Bacon.

Lawler, E. J., R. Ford, and M. A. Blegen. 1988. "Coercive Capability in Conflict: A Test of Bilateral Deterrence vs. Conflict Spiral Theory." *Social Psychology Quarterly* 51: 93-107.

Lawler, E. J., C. Ridgeway, and B. Markovsky. 1993. "Structural Social Psychology and the Micro-Macro Problem." *Sociological Theory* 11: 268-290.

Nemeth, C. 1972. "A Critical Analysis of Research Utilizing the Prisoner's Dilemma Paradigm for the Study of Bargaining." *Advances in Experimental Social Psychology* 6: 203-34.

Pruitt, D. G. 1981. *Negotiation Behavior*. New York: Academic Press.

Pruitt, D. G., and M. J. Kimmel. 1977. " Twenty Years of Experimental Gaming: Critique, Synthesis, and Suggestions for the Future." *Annual Review of Psychology* 28: 363-392.

Rapoport, A. 1966. *Two Person Game Theory: The Essential Ideas*. Ann Arbor: University of Michigan Press.

Rubin, J. Z., and B. R. Brown. 1975. *The Social Psychology of Bargaining and Negotiation*. New York: Academic Press.

Secord, P. 1977. "Social Psychology in Search of a Paradigm." *Personality and Social Psychology Bulletin* 3: 41-50.

Schelling, T. 1960. *The Strategy of Conflict*. New York: Oxford University Press.

Schlenker, B., and T. V. Bonoma. 1978. "Fun and Games: The Validity of Games for the Study of Social Conflict." *Journal of Conflict Resolution* 22: 7-38.

Stryker, S. 1977. "Developments in 'Two Social Psychologies': Toward an Appreciation of Mutual Relevance." *Sociometry* 40: 145-160.

_____. 1989. "The Two Psychologies: Additional Thoughts." *Social Forces* 68: 45-54.

Tedeschi, J., T. V. Bonoma, and R. C. Brown. 1971. "A Paradigm for the Study of Coercive Power." *Journal of Conflict Resolution* 15: 198-222.

Tedeschi, J., B. R. Schlenker, and T. V. Bonoma. 1973. *Conflict, Power, and Games*. Chicago: Aldine.

Wagner, D. G. 1984. *The Growth of Sociological Theories*. Beverly Hills, CA: Sage.

Wagner, D. G., and J. Berger. 1985. "Do Sociological Theories Grow?" *American Journal of Sociology* 90: 697-728.

Walker, H. A., and Bernard P. Cohen. 1985. "Scope Statements: Imperatives for Evaluating Theory." *American Sociological Review* 50: 288-301.

Webster, M., Jr. 1994. "Experimental Methods." Pp. 43-69 in *Group Processes: Sociological Analyses*, edited by M. Foschi and E. J. Lawler. Chicago: Nelson Hall.

Webster, M., Jr., and J. B. Kervin. 1971. "Artificiality in Experimental Sociology." *Canadian Review of Sociology and Anthropology* 8: 263-272.

Willer, D. 1987. *Theory and the Experimental Investigation of Social Structures.* New York: Gordon and Breach Science Publishers.

Zelditch, M., Jr. 1968. "Can You Really Study an Army in the Laboratory?" Pp 123-145 in *Complex Organizations: A Sociological Reader,* edited by A. Etzioni. New York: Holt, Rinehart & Winston.

J A I   P R E S S

Advances in Group Processes

Edited by **Edward J. Lawler**, *Department of Organizational Behavior, Cornell University*

REVIEWS: "A major impression one gets from this volume is that far from being dormant, the social psychology of groups and interpersonal relations is quite vibrant, and very much involved with compelling problems."

"Concerns about the imminent demise of group processes as an area of study in social psychology are clearly exaggerated. But should doubts remain, they ought to be allayed by the range and quality of the offerings in this volume."

" . . . should be of interest both to specialists in group processes and to sociologists who are interested in theory, particularly in theoretical linkages between micro and macro analysis. Because many of the papers offer thorough reviews and analyses of existing theoretical work, as well as new theoretical ideas, they are also useful readings for graduate students."

— *Contemporary Sociology*

Volume 12, 1995, 300 pp. $73.25
ISBN 1-55938-872-2

CONTENTS: Preface, *Barry Markovsky, Karen Heimer and Jodi O'Brien.* Social Influence and Religious Pluralism, *William Sims Bainbridge.* Gender Differences in Reference Groups, Self-Evaluations, and Emotional Experience Among Employed Married Parents, *Robin W. Simon.* Modeling Individual Perceptions of Justice in Stratification Systems, *Yuriko Saito.* What Does it Mean Social Psychologically to be of a Given Age, Sex-Gender, Social Class, Race, Religion, Etc?, *Theodore D. Kemper.* Sexual Orientation as a Diffuse Status Characteristic: Implications for Small Group Interaction, *Cathryn Johnson.* The Determinants of Top Management Teams, *Karen A. Bantel and Sydney Finkelstein.* Social Determinants of Creativity: Status Expectations and the Evaluation of Original Products, *Joseph Kasof.* Role as Resource in a Prison Uprising, *Peter L. Callero.* Social Identification and Solidarity: A Reformulation, *Barry Markovsky and Mark Chaffee.* Mutual Dependence and Gift Giving in Exchange Relations, *Edward J. Lawler, Jeongkoo Yoon, Mouraine R. Baker and Michael D. Large.*

Also Available:
Volumes 1-11 (1984-1994) $73.25 each

Advances in Managerial Cognition and Organizational Information Processing

Edited by **James R. Meindi,**
State University of New York at Buffalo

Volume 5, 1994, 337 pp. $73.25
ISBN I-55938-447-6

Edited by **Chuck Stubbart,** *Southern Illinois University,*
James R. Meindi, *State University of New York
at Buffalo* and **Joseph F. Porac,** *University of Illinois*

CONTENTS: Introduction, *Chuck Stubbart, James R. Meindi
and Joseph F. Porac.* PART I: THEORETICAL ARTICLES.
Toward a Cognitive Theory of Organizations, *Mariann Jelinek
and Joseph A. Litterer.* Mapping Learning Processes in Organizations: A Multi-Level Framework Linking Learning and Organizing, *Mary Ann Glynn, Theresa K. Lant and Frances J.
Milliken.* A Theory of Temporal Adjustments of the Evaluation
of Events: Rosy Prospection and Rosy Retrospection, *Ter-
ence R. Mitchell* and *Leigh Thompson.* Performance Appraisal: The Influence of Liking on Cognition, *Robert L. Cardy* and
Gregory H. Dobbins. Techniques to Compare Cognitive
Maps, *Kevin Daniels, Livia Markoczy,* and *Leslie de Chem-
atony.* Do Cognitive Maps Make Sense? *W. J. Scheper* and *J.
Faber.* Negotiation As Problem Solving, *Michael J. Prietula*
and *Laurie R. Weingart.* Computers and Competitive Advantage — The Strategic Management of Information Systems,
Ewan Sutherland. PART II: EMPIRICAL ARTICLES. Interpret-
ing Strategic Issues: Making Sense of "1992", *Susan C.
Schneider.* Modeling Individual Knowledge in the Personnel
Evaluation Process, *Guido Capaldo* and *Giuseppe Zollo.*
Feedback Sign and "Mindful Vs. Mindless" Information Pro-
cessing, *Kenneth J. Dunegan.* Computer-Aided Text Analysis:
Are Words A Window To The Mind? *Beverly C. Winterscheid.*

Also Available:
Volumes 1-4 (1984-1994) $73.25 each

J
A
I

P
R
E
S
S

Advances in Interdisciplinary Studies of Work Teams

Edited by **Michael Beyerlein,** *Director,*
Interdisciplinary Center For The Study of
Work Teams, University of North Texas

Volume 2, Knowledge, Work in Teams
1995, 301 pp. $73.25
ISBN 1-55938-926-5

Edited by **Michael M. Beyerlein, Douglas A.**
Johnson, *Center for the Study of Work Teams,*
Department of Psychology, University of North Texas
and **Susan T. Beyerlein,** *Center for Public Management,*
University of North Texas

Also Available:
Volume 1 (1994) $73.25

**J
A
I
P
R
E
S
S**